Lecture Notes in Mathematics   1640

Editors:
A. Dold, Heidelberg
F. Takens, Groningen

Subseries: Fondazione C.I.M.E.
Advisor: Roberto Conti

Springer
Berlin
Heidelberg
New York
Barcelona
Budapest
Hong Kong
London
Milan
Paris
Santa Clara
Singapore
Tokyo

G. Bolliat   C.M. Dafermos
P.D. Lax   T.P. Liu

Recent Mathematical Methods in Nonlinear Wave Propagation

Lectures given at the 1st Session of the
Centro Internazionale Matematico Estivo
(C.I.M.E.), held in Montecatini Terme, Italy,
May 23–31, 1994

Editor: T. Ruggeri

Fondazione
C.I.M.E.

 Springer

Authors

Guy Boillat
Lab. de Recherches Scientifiques
et Techniques
Université de Clermont-Ferrand II
63177 Aubière Cedex, France

Tai-Ping Liu
Dept. of Mathematics
Stanford University
Stanford, CA 94305-2125, USA

Constantin M. Dafermos
Applied Mathematics Division
Brown University
Providence, RI 02912, USA

Editor

Peter D. Lax
Courant Institute
New York University
251 Mercer Street
New York, NY 10012, USA

Tommaso Ruggeri
C.I.R.A.M.
Università di Bologna
Via Saragozza, 8
I-40126 Bologna, Italy

Die Deutsche Bibliothek – CIP-Einheitsaufnahme

Centro Internazionale Matematico Estivo <Firenze>:
Lectures given at the . . . session of the Centro Internazionale Matematico Estivo (CIME) . . . – Berlin;
Heidelberg; New York; London; Paris; Tokyo; Hong Kong: Springer
Früher Schriftenreihe. – Früher angezeigt u.d.T.: Centro Internazionale Matematico Estivo: Proceedings of
the session of the Centro Internazionale Matematico Estivo (CIME)
NE: HST   1994,1. Recent mathematical methods in nonlinear wave propagation. – 1996

Recent mathematical methods in nonlinear wave propagation:
held in Montecatini Terme, Italy, May 23–31, 1994 / G. Boillat . . . Ed.: T. Ruggeri. – Berlin; Heidelberg;
New York; Barcelona; Budapest; Hong Kong; London; Milan; Paris; Santa Clara; Singapore; Tokyo:
Springer, 1996
(Lectures given at the . . . session of the Centro Internazionale Matematico Estivo (CIME) . . . ; 1994,1)
(Lecture notes in mathematics; Vol. 1640: Subseries: Fondazione CIME)
ISBN 3-540-61907-0
NE: Boillat, Guy; Ruggeri, Tommaso [Hrsg.]; 2. GT

Mathematics Subject Classification (1991):
35L, 35M. 35Q, 76N, 76L05, 76P05, 76W05, 76Y05, 78A25

ISSN 0075-8434
ISBN 3-540-61907-0 Springer-Verlag Berlin Heidelberg New York

© Springer-Verlag Berlin Heidelberg 1996
Printed in Germany

The use of general descriptive names, registered names, trademarks, etc. in this publica-
tion does not imply, even in the absence of a specific statement, that such names are
exempt from the relevant protective laws and regulations and therefore free for general
use.

Typesetting: Camera-ready TEX output by the authors
SPIN: 10479918   46/3142-543210 - Printed on acid-free paper

PREFACE

The book contains the text of the lectures presented at the first session of the Summer School 1994 organized in Montecatini Terme by the C.I.M.E. Foundation.

The aim of the School was the presentation of the state of the art on recent mathematical methods arising in Nonlinear Wave Propagation.

The lecture notes presented in this volume were delivered by leading scientists in these areas and deal with *Nonlinear Hyperbolic Fields and Waves* (by Professor G. Boillat of Clermont University), *The Theory of Hyperbolic Conservation Laws* (by Professor C. M. Dafermos of Brown University), *Outline of a Theory of the KdV Equation* (by Professor P. D. Lax of Courant Institute NYU), *Nonlinear Waves for Quasilinear-Hyperbolic-Parabolic Partial Differential Equations* (by Professor T.-P. Liu of Stanford University).

About fifty people (including research students and senior scientists) participated actively in the course. There were also several interesting contributions from the seminars on specialized topics.

We feel that the volume gives a coherent picture of this fascinating field of Applied Mathematics.

Tommaso Ruggeri

Contents

Non Linear Hyperbolic Fields and Waves

Guy Boillat

En face de la nature, il faut étudier toujours,

mais à la condition de ne savoir jamais.

R. Töpffer

Introduction

Nonlinearity and hyperbolicity are essential features of Mechanics and Relativity. In the last decennials much work has been done leading to a better understanding of the systems in conservative form. Physics, however, as Infeld and Rowlands remark[0], still widely ignores the interesting properties of nonlinear theories. The topics presented in these lectures with physical examples include: discontinuity waves and shocks with particular emphasis on exceptional waves and characteristic shocks; symmetrization of conservative systems compatible with an entropy law, subluminal velocities in relativistic theories, systems with involutive constraints, new field equations by means of generators with special attention to extended thermodynamics and nonlinear electrodynamics. It is our hope that the applications proposed in these lectures will awake a large interest in the nonlinearity of Nature.

1. Hyperbolicity, Conservative form

The N components of the column vector $u(k^\alpha)$ $(\alpha = 0, 1, ..., n)$ satisfy the N partial differential equations of the quasi-linear system

$$(1) \qquad A^\alpha(u)u_\alpha = f(u), \quad u_\alpha = \partial u/\partial x^\alpha.$$

The matrix A^0 is supposed to be regular so that the system is usally written

$$u_t + A^i(u)u_i = f(u), \quad x^0 = t, \quad i = 1, 2, ..., n.$$

It is hyperbolic if, for any space vector $\vec{n} = (n_i)$ the matrix $A_n := A^i n_i$ has a complete set of (i.e., N) real eigenvectors [1]-[8]. If all eigenvalues are distinct hyperbolicity is *strict*. In three-dimensional space this will not be possible for all $\vec{n}$ when $N = \pm 2, \pm 3, \pm 4$ (mod 8) [9],[10]. Therefore, there must correspond $m^{(j)}$ eigenvectors to the eigenvalue $\lambda^{(j)}$ of multiplicity $m^{(j)}$. We denote them with the initials of the Latin words *laevus* (left) and *dexter* (right). Two indices are needed, one for the eigenvalue, the other one for the multiplicity : $d_J^{(j)}$. However, the subscript suffices if we agree that $d_J, d_{J'}, (J, J' = 1, 2, ...m^{(j)})$ are eigenvectors corresponding to $\lambda^{(j)}, d_K, d_{K'}$ to $\lambda^{(k)}$ etc. Hence

$$(2) \qquad \ell_J(A_n - \lambda^{(j)}A^0) = 0, \quad (A_n - \lambda^{(k)}A^0)\, d_K = 0.$$

It can also be assumed that $\ell_J A^0 d_{J'} = \delta_{JJ'}$.

For the system of *balance laws* encountered in Mechanics and Physics

$$(3) \qquad \partial_\alpha f^\alpha(u) = f(u)$$

the matrices A^α are equal to the gradient of the vectors f^α with respect to u

$$(4) \qquad A^\alpha = \nabla f^\alpha,$$

so that with the choice $u := f^0, A^0 = I$. Although (3) expresses *conservation laws* only when there is no second member, this system is nevertheless said to have *conservative form* in reference to its first member.

2. Wave velocities

Discontinuities $[u_\alpha]$ of the first order derivatives u_α may occur across some characteristic surface (wave front) $\varphi(x^\alpha) = 0$. In fact by a well known result of Hadamard [11]

$$[u_\alpha] = \varphi_\alpha \delta u$$

and taking the jump of (1) results in

$$A^\alpha \varphi_\alpha \delta u = 0$$

or, by introducing *normal wave velocity* $\lambda \vec{n}$

$$(5) \qquad \lambda = -\varphi_t/|\text{grad } \varphi|, \quad n_i = \varphi_i/|\text{grad } \varphi|$$

$$(A_n - \lambda I)\,\delta u = 0.$$

The physical meaning of the eigenvalue $\lambda^{(i)}$ is thus quite simple : it is the velocity of the wave front propagating the weak discontinuity

$$(6) \qquad \delta u = \pi^I d_I, \quad I = 1, 2, ..., m^{(i)}.$$

It was Courant who suggested to Peter Lax that he study the evolution of these discontinuities [12] and they showed [1],[13] how they propagate along the bicharacteristic curves

$$(7) \qquad dx^\alpha/d\sigma = \partial\psi^{(i)}/\partial\varphi_\alpha, \quad d\varphi_\alpha/d\sigma = -\partial\psi^{(i)}/\partial x^\alpha,$$

$$(8) \qquad \psi^{(i)} := \varphi_t + |\text{grad } \varphi|\, \lambda^{(i)}(u, \vec{n}) = 0.$$

The velocity of propagation $\partial\psi/\partial\varphi_i$ is the *ray velocity*

$$(9) \qquad \vec{\Lambda} = \lambda\,\vec{n} + \partial\lambda/\partial\vec{n} - \vec{n}(\vec{n}.\partial\lambda/\partial\vec{n}), \quad \lambda = \vec{\Lambda}.\vec{n}.$$

In the non-linear case the components π^I satisfy along the rays a Bernoulli system of differential equation [44], [14]

$$(10) \qquad d\pi^I/d\sigma + a^I_{I'}\pi^{I'} + |\text{grad } \varphi|\, \pi^{I'}\,\pi^I\,\nabla\lambda^{(i)}d_{I'} = 0$$

where the coefficients depend on the solution in the unperturbed state u_0 and on the geometry of the wave front. (Another approach [15] involving discontinuities in the derivatives of φ, leads to an equivalent though different system [16]).

For *asymptotic waves* [1], [17], [18]

$$u = u_0(x^\alpha) + u_1(x^\alpha; \xi)/\omega + u_2(x^\alpha; \xi)/\omega^2 + \dots, \quad \xi = \omega\varphi$$

the equations of evolution given by Y. Choquet-Bruhat [19]-[22]

$$\text{(11)} \qquad \partial u^I/\partial\sigma + |\text{grad } \varphi| \, u^{I'} (\nabla\lambda^{(i)} d_{I'}) \, \partial u^I/\partial\xi + a^I_{I'} u^{I'} = 0, \; u_1 = u^I d_I$$

yields also (10) as a special solution : $u^I = \xi \, \pi^I$. When u_0 depends on ξ see D. Serre [23].

When the disturbance propagates into a constant state u_0 equation (7) shows that the points M of the wave front S at time σ are related to those M_0 of the initial wave surface S_0 by

$$\vec{M}(\sigma) = \vec{M}_0 + \vec{\Lambda}(u_0, \vec{n}_0) \, \sigma, \quad \vec{n}(M) = \vec{n}(M_0) = \vec{n}_0.$$

It follows that S and S_0 are obtained by translation (or are parallel surfaces) if $\vec{\Lambda}$ (or λ) do not depend on $\vec{n}$.

In the absence of a source term a *simple wave* [24]-[26] solution $u = u(\varphi)$ satisfies the ordinary differential system [27]

$$\text{(12)} \qquad du/d\varphi = \alpha^J(\varphi) \, d_J(u, \vec{n})$$

where $\varphi(t, x^i)$ is explicitly defined by

$$g(\varphi) = x^i n_i - \lambda^{(j)}(u, \vec{n})t, \quad \vec{n} = \text{const},$$

and $g = \varphi$ or 0 (for centred waves).

This simple wave solution is important because it describes the state adjacent to a constant state [26]. It singles out among the solutions for which the direction of u_x is submitted to some restrictions [28].

The velocity varies according to

$$\text{(13)} \qquad d\lambda^{(j)}/d\varphi = \alpha^J \nabla\lambda^{(j)} d_J.$$

The characteristic equations of covariant field equations appear in covariant form as

$$\text{(14)} \qquad \psi := G^{\alpha\beta\dots\gamma}\varphi_\alpha\varphi_\beta\dots\varphi_\gamma = 0$$

where G is a completely symmetric tensor. The ray velocity is given by

$$\text{(15)} \qquad \Lambda^\alpha = \partial\psi/\partial\varphi_\alpha$$

and, in a relativistic theory, must not exceed the velocity of light i.e., the ray velocity must be a time-like (or null) vector

$$\text{(16)} \qquad g_{\alpha\beta}\Lambda^\alpha\Lambda^\beta \geq 0$$

while the wave surface, by $\Lambda^\alpha\varphi_\alpha = 0$, satisfies

$$g^{\alpha\beta}\varphi_\alpha\varphi_\beta \leq 0.$$

The Rarita-Schwinger wave fronts, on the contrary, propagate faster than light [29].

3. Exceptional waves

The evolution equations (10) and (11) are nonlinear unless

(17) $$\nabla\lambda^{(i)}d_I \equiv 0, \quad I = 1, 2, ..., m^{(i)}.$$

In this case we say with Lax [25], [26] that the wave is *exceptional*, for there is no reason, in general, for the gradient of $\lambda^{(i)}$ to be orthogonal to its eigenvectors. However, since it is clearly the case for linear fields ($\nabla\lambda^{(i)} \equiv 0$), the velocity is also said to be *linearly degenerated*. Instead, when $\nabla\lambda.d \neq 0$ the characteristic field is *genuinely nonlinear*. By (6) the condition (17) is simply written

(18) $$\delta\lambda^{(i)} \equiv 0$$

or by (9)

(19) $$\vec{n}.\delta\vec{\Lambda}^{(i)} = 0.$$

The disturbance of the ray velocity, the only physically meaningful vector, one can derive from (1), without further information on the field equations, is orthogonal to the wave normal. Therefore, such a wave may also be called *transverse wave* [2].

The corresponding simple wave, by (13), moves with a constant velocity and first integrals of (12) can then easily be found for conservation laws. In fact, since

$$d(f_n - \lambda^{(j)}u)/d\varphi + ud\lambda^{(j)}/d\varphi = 0$$
$$f_n(u) - \lambda^{(j)}(u)u = f_n(u_0) - \lambda^{(j)}(u_0)\, u_0, \quad \lambda^{(j)}(u) = \lambda^{(j)}(u_0), \quad u_0 = u(0).$$

The explicit solution of these equations will be given below [30].

In a covariant formulation exceptionality of (14) is expressed by

(20) $$\delta\psi = \varphi_\alpha\varphi_\beta...\varphi_\gamma\delta G^{\alpha\beta...\gamma} = 0, \quad \varphi_\alpha\delta\Lambda^\alpha = 0.$$

Only in this case does the disturbance of a tensor depending on φ_α have a covariant meaning [31].

Actually, in spite of their name, exceptional waves are rather common and can be encountered for instance, in the equations of Einstein for gravitation, of the fluids (entropy and Alfvén waves [2]), of Monge-Ampère, of nonlinear electromagnetism, of Nambu [32]. Also *multiple waves of conservative systems are exceptional* [33], [34]. To see this take the derivative of (2) in the direction of $d_{K'}$,

$$\left(\nabla A_n - \lambda^{(k)}\nabla A^0\right) d_{K'}d_K + \left(A_n - \lambda^{(j)}A^0\right)\nabla d_K d_{K'} - A^0 d_K(\nabla\lambda^{(k)}d_{K'}) = 0,$$

exchange K and K' and substract. The first terms drop out, due to (4), and the result follows. As a consequence the system of two equations

$$u_t + w^i(u, v)\, u_i = 0, \quad v_t + w^i(u, v)\, v_i = 0$$

which has the double velocity $\lambda = w^i n_i$ cannot be put in a conservative form unless the w^i's are constant (hence λ exceptional).

Incidentally, when two velocities crossed for some value $\vec{n}_0(u)$ of $\vec{n}$, a possibility evoked in the first paragraph, thus creating a variable multiplicity, the important criterion is the exceptionality (for $\vec{n}_0$) of the difference of these two velocities [35].

As an illustration consider the Euler equations of a fluid

$$\partial_t \rho + \operatorname{div}(\rho \vec{v}) = 0$$

(21)
$$\partial_t(\rho \vec{v}) + \partial_i(\rho \, v^i \vec{v}) + \overrightarrow{\operatorname{grad}} \, p = 0$$

$$\partial_t(\rho S) + \operatorname{div}(\rho S \vec{v}) = 0.$$

To compute the perturbations and velocities one makes the substitutions

(22)
$$\partial_t \to -\lambda \delta, \ \partial_i \to n_i \delta, \ \overrightarrow{\operatorname{grad}} \to \vec{n} \delta, \ \operatorname{curl} \to \vec{n} \times \delta$$

or, simply in a covariant theory

(23)
$$\partial_\alpha, \ \nabla_\alpha \longrightarrow \varphi_\alpha \delta.$$

Here one immediately obtains, with $p = p(\rho, S)$,

$$(v_n - \lambda) \, \delta \rho + \rho \delta v_n = 0$$

$$\rho(v_n - \lambda) \, \delta \vec{v} + \vec{n} \delta p = 0$$

$$(v_n - \lambda) \, \delta S = 0.$$

The entropy wave ($\delta S \neq 0$) moves with the fluid, $\lambda = v_n$, is exceptional, $\delta v_n = \delta p = 0$ and has a multiplicity equal to three (the number of arbitrary disturbances). The ray velocity is $\vec{v}$ and $\delta \vec{v} \neq 0$.

For the sonic wave $\delta S = 0, \delta p = p' \delta \rho, \ \lambda = v_n \pm \sqrt{p'}$. The ray velocity is $\vec{\Lambda} = \vec{v} \pm \sqrt{p'} \, \vec{n}$.

Further

(24)
$$\delta \lambda = \vec{n}.\delta \vec{\Lambda} = \pm \sqrt{p'} \left(\frac{1}{\rho} + \frac{p''}{2p'} \right) \delta \rho.$$

The nonlinear d'Alembert equation derives from a variational principle with the Lagrangian [46]

$$L = L(Q), \ Q = -\frac{1}{2} \, g^{\alpha\beta} \partial_\alpha \, u \partial_\beta u, \ g^{\alpha\beta} = \operatorname{diag}(1, -1, -1, -1)$$

and reads

(25)
$$\partial_\alpha \, (L' u^\alpha) = 0, \ (L' \, g^{\alpha\beta} - L'' \, u^\alpha u^\beta) \, \partial_{\alpha\beta} u = 0, \ u^\alpha = g^{\alpha\beta} \, \partial_\beta u.$$

With the substitution (23) the wave front is easily obtained

$$\psi = L' \, g^{\alpha\beta} \varphi_\alpha \varphi_\beta - L'' \, (u^\alpha \varphi_\alpha)^2 = 0$$

since

$$\delta u_\alpha = \varphi_\alpha \delta^2 u.$$

According to (15) the ray velocity is

$$\Lambda^\alpha = L' g^{\alpha\beta} \varphi_\beta - L''(u^\beta \varphi_\beta) u^\alpha$$

and, by (16),

(26) $$g_{\alpha\beta} \Lambda^\alpha \Lambda^\beta = -(u^\alpha \varphi_\alpha)^2 (L' + 2QL'') L'' \geq 0.$$

The expression of its covariant perturbation

(27) $$\delta \Lambda^\alpha / \delta^2 u = -2(u^\beta \varphi_\beta) L'' \Lambda^\alpha / L' + (u^\beta \varphi_\beta)^2 u^\alpha (L''' - 3L''^2/L')$$

is valid, as mentioned above, only in the exceptional case when $\varphi_\alpha \delta \Lambda^\alpha = 0$ i.e.,

(28) $$L''' - 3L''^2/L' = 0$$

and reduces to its first term. However, it is true, in general, that

(29) $$\varphi_\alpha \delta \Lambda^\alpha = (u^\alpha \varphi_\alpha)^3 (L''' - 3L''^2/L') \delta^2 u.$$

Now appears the difference with (24). For a fluid the scalar $\nabla \lambda d$ does not change sign when $\vec{n}$ varies (cf. the cases of magnetohydrodynamics [44] and nonlinear electromagnetic media) [45]. Therefore, a spherical perturbation propagating in a constant state will either increase without limit or decrease (compression or rarefaction wave). On the contrary, if $Q > 0$ (29) changes sign across the critical cone [36]

$$\left| \frac{u_i}{u_0} n_i \right| = 1$$

and the behaviour of the perturbation will depend on the direction of propagation.

In one-dimensional propagation directional exceptionality cannot occur but it may happen that the condition of linear degeneracy be verified only for some value of the field u. This phenomenon appears in a rigid heat conductor and has been studied in detail by Ruggeri, Muracchini and Seccia. They find critical temperatures in agreement with those experimentally observed for the *NaF* and *Bi* crystals [42].

Of interest in microphysics [37] the scalar field also describes the static irrotational isentropic and supersonic flow. In this case ([24], p.26 sqq., [38], p.201)]

$$\vec{v} = \text{grad } u$$

and (25) can be rewritten

$$(c^2 \delta_{ij} - u_i u_j) \partial_{ij} u = 0, \quad c^2(Q) = -L'/L'', \quad Q = \frac{1}{2} \vec{v}^2.$$

Although there is no velocity of propagation this equation is still hyperbolic provided that $v^2 > c^2$. This is the condition for the existence of the characteristic surface whose normal satisfies

$$(\vec{v}.\vec{n})^2 = c^2$$

implying

(30) $$(L' + 2QL'')L'' > 0.$$

4. Exceptionality as a principle of selection

The concept of linear degeneracy introduced by Peter Lax plays a fondamental role in the propagation of weak and, as we shall see later, of strong discontinuities. Therefore, when the field equations are not completely determined one can require the exceptionality of some wave(s) to solve the indeterminacy and then inquire about the result which is expected to have some special physical or mathematical meaning.

A) Fluid

For a fluid equation (24) yields immediately the equation of state

(31) $$p = b(S) - a(S)/\rho$$

which is the well-known law of von Kármán-Tsien and frequently approximates the pressure in the theory of subsonic flow ([24], p.10), [40]). A similar law

$$p = b - a^2/(\rho + bc^{-2})$$

holds in the relativistic case (ρ is now the density of energy) but another solution exists [41]

$$p = \rho c^2$$

the equation of state of the incompressible fluid whose sound waves travel at the limit [90] speed of light c both solutions of [19], [31], [87]

(32) $$(\rho + p/c^2)\, p'' + 2p'\,(1 - p'/c^2) = 0, \quad p' = \partial p\,(\rho, S)/\partial \rho.$$

This last solution is the only possible one for a completely exceptional charged fluid [52].

B) Elastic tube

The velocity v of a fluid of constant density ρ filling an elastic tube of section s satisfies the equations

$$\partial_t s + \partial_x(sv) = 0$$

$$\partial_t v + \partial_x(\frac{1}{2}\,v^2 + p/\rho) = 0, \quad p = p(s).$$

We see immediately that the wave speeds are

$$\lambda = v \pm \sqrt{sp'/\rho}$$

and

$$\delta\lambda = (\lambda - v)\,(\frac{1}{s} + \frac{1}{2}\,\frac{(sp')'}{sp'})\,\delta s.$$

Thus the exceptional case corresponds to the pressure

$$p(s) = -a/2s^2 + b$$

which is the law of a rubber-like material. The theoretical and experimental study shows indeed that no shock forms [59].

C) Scalar field

Instead the scalar field will be completely exceptional if (29) vanishes i.e., if the Lagrangian satisfies

$$L'L''' - 3L''^2 = 0$$

the nonlinear solution of which is

$$L = \sqrt{2k_1 Q + k_2} + k_3$$

with two superfluous constants. Since, by (26), k_1 and k_2 are of the same sign one can choose

$$(33) \qquad L = \sqrt{2Q + k}, \quad k > 0.$$

This Lagrangian introduced by Max Born [47] has also been considered by Heisenberg [48]. In one space dimension the characteristic curves are isocline [46], ([38], p.579, 617-20), [49].

From (27) follows

$$\delta \Lambda^{\alpha} // \Lambda^{\alpha}.$$

so that this weak discontinuity is zero when the ray velocity is normalized in time ($\Lambda^0 = 1$) or in space-time ($g_{\alpha\beta}\Lambda^{\alpha}\Lambda^{\beta} = 1$).

This is general : *the (normalized) ray velocity is not disturbed for an exceptional wave of Euler's variational equations (with a convex density of energy)* [131].

We have just seen for the entropy wave that

$$\delta\lambda = \delta v_n = 0, \ \delta\vec{\Lambda} = \delta\vec{v} \neq 0.$$

Hence, the fluid equations do not derive (at least in three dimensions) from a unique (Lagrangian) function but as we shall see below from a potential vector.

On the other hand, the elliptic signature of the metric for the static irrotational fluid implies with (30), (33) $k < 0$ and

$$c^2 = v^2 - |k|.$$

The analogy between the von Kármán fluid and the Born theory has been observed very early by means of the hodograph transformation [50], [51].

D) Born-Infeld Theory

If we apply a variational principle to a Lagrangian depending on the electromagnetic invariants [53]-[55]

$$Q = \frac{1}{4} F_{\alpha\beta} F^{\alpha\beta}, \ R = \frac{1}{4} F_{\alpha\beta} \overset{*}{F}{}^{\alpha\beta}$$

the corresponding Euler equations

$$(34) \qquad \partial_{\alpha}(L_Q F^{\alpha\beta} + L_R \overset{*}{F}{}^{\alpha\beta}) = 0, \ F_{\alpha\beta} = \partial_{\alpha}\phi_{\beta} - \partial_{\beta}\phi_{\alpha}$$

yield the Maxwell equations when L is a linear function of Q and R with the result that the electric field of a spherically symmetric particle at rest decreases like the inverse square of the distance meaning also that the field grows without limit when the distance tends to zero. By taking a non-linear Lagrangian, Born and Infeld solved this difficulty and did much more. Max Born has told how the idea came [56] ; it all began in 1933 in Selva/Wolkenstein, South Tyrol...

There are two families of wave fronts satisfying the characteristic equation [57]

$$(\tau^{\alpha\beta} + \mu g^{\alpha\beta}) \varphi_{\alpha}\varphi_{\beta} = 0$$

where $\tau^{\alpha\beta}$ is the usual Maxwell tensor and $\mu(Q, R)$ takes on two values μ_1 and μ_2 determined by the knowledge of the Lagrangian.

It turns out that these two values coincide only when

(35) $$L = \sqrt{-R^2 + k(2Q + k)}, \quad \mu = Q + k$$

which is just the Born-Infeld Lagrangian. Due to the multiplicity of the wave the related system (34) is therefore completely exceptional but it is not the only one and other Lagrangians share this property with (35). They have been determined. The general one depends on several constants and gives (35) when one of them, maybe related to the Planck constant, vanishes [57].

At distance r the electric field of a charged particle is just

$$E(r) = \sqrt{k}/\sqrt{1 + (r/r_0)^4}, \quad E(0) = \sqrt{k}$$

so that $E^2 \leq k$. More generally, for an admissible Lagrangian,

$$E^2 \leq \min(\zeta_1, \zeta_2), \quad \zeta = \mu - Q$$

and the limit value ζ is called the absolute field. The electric field reaches this limit in the frame moving with the ray velocities. For this reason, having also in mind the entropy wave of continuum mechanics, we proposed [57] :

A stable particle moves along an exceptional bicharacteristic.

Thus are determined the equations of motion of the particles in nonlinear electrodynamics. The Born-Infeld theory was the first attempt to deal with the difficulties of microphysics by means of nonlinear equations. Although its development was hindered by this very nonlinearity it reveals from the mathematical point of view an interesting structure. From the physical point of view it has been shown that the present relativistic strings and membranes are just particular solutions of Born-Infeld [32], [58].

E) Elasticity

Some waves propagating in elastic solids may be naturally exceptional at least in certain direction or for certain kind of deformation [60]-[62]. Instead the requirement of linear degeneracy [63]-[66] leads to the determination of classes of elastic potentials containing the potentials of Grioli [67], Hadamard, Hooke, Mooney-Rivlin and Tolotti [68].

F) Monge-Ampère equation

The nonlinear equation

$$u_{tt} + f(x, t, u, p, q, r, s) = 0$$
$$p = u_x, \quad q = u_t, \quad r = u_{xx}, \quad s = u_{xt}$$

becomes quasi-linear when derivated with respect to t.

Assuming therefore discontinuities of the third order across a wave surface the application of the operator (22) gives $\delta r = \delta^3 u$, $\delta s = -\lambda\delta^3 u$

$$\lambda^2 - \lambda f_s + f_r = 0.$$

The requirement that $\delta\lambda$ be zero for both velocities leads to

$$f_{rr} - f_r f_{ss} = 0, \quad f_s f_{ss} - 2f_{sr} = 0$$

the integration of which results in [69]

$$F := H u_{tt} + 2K u_{tx} + L u_{xx} + M + N \left(u_{tt} u_{xx} - u_{tx}^2\right) = 0,$$

an equation due to Monge and Ampère (when $N \neq 0$) [70] and well suited for initial-value problems ([1], p.499).

Now it is natural to use this characteristic property to extend its form to n independent variables or to higher orders. It follows in the first case that the Monge-Ampère function F *is a linear function of the Hessian of u and of all its minors* [71] in accordance with the results found by direct calculation for $n = 3$ [72] or $n = 4$ [73]. In the second case introducing

$$X_k = \partial^n u / \partial t^k\, \partial x^{n-k}, \quad 0 \leq k \leq n$$

and the Hankel matrix $K(K_{ij} = X_{i+j}; i, j = 0, 1, 2, ...m)$ *the Monge-Ampère function F is a linear function of all minors of K (including K) if n = 2m or of K without its last row if n = 2m − 1* [74].

The Natan equation [55]

$$\sum_{i,j} b_{ij} \left(X_i X_{j+1} - X_j X_{i+1}\right) + \sum_0^n a_i X_i + a = 0$$

is of this type.

A long-forgotten equation hundreds of papers have appeared in the last years on the Monge-Ampère field. It is also applied. For instance, a one-dimensional von Kármán fluid is such a field [75] and the third order equation has recently been used for the thermodynamics of fluctuation [76], [77]. Sometimes the classical equation can even be explicitly integrated (generalized Born-Infeld Lagrangians are Monge-Ampère solutions [57]) and this is not the least interesting feature of this equation.

All these equations belong to the general class of the Monge-Ampère systems which, among the nonlinear partial differential equations, is the closest to the quasi-linear class sharing with it an important linear property [78].

5. Symmetric systems. Symmetrization.

If the matrices A^α are symmetric and if A^0 is positive definite the system (1) is called symmetric [1], [79]. It is always hyperbolic and the Cauchy problem is well-posed [80], [81]. But where are these systems to be found? In 1961 Godunov [82] discovered that the Euler equations of the fluid where the conservation of energy

$$\partial_t \left\{\rho(e + \frac{1}{2}\, v^2)\right\} + \partial_i \left\{\rho(e + \frac{1}{2}\, v^2 + p/\rho)\, v^i\right\} = 0$$

replaces the last equation (21) were part of this class as well as the Euler variational equations with the Lagrangian $L(\partial_\alpha q^s)$.

In fact introducing the new field variables

(36) $\check{u}' = \dfrac{1}{T}\,(G - \dfrac{1}{2}\,\vec{v}^2, \vec{v}, -1),\; G = e + pV - TS,\; de = TdS - pdV,\quad V = 1/\rho,$

(e is the internal energy and G the chemical potential) it appears that $f^0 = u$ and f^i are just the derivatives with respect to u' of the functions

$$h'^0 = p/T,\; h'^i = pv^i/T;\; f^\alpha = \check{\nabla}' h'^\alpha$$

so that (3) takes a special symmetric form. The same can be done for the variational equations (see below). Godunov then deduced for systems of this form the existence of an additional conservation law.

Precisely in mechanics and physics systems of balance laws are supplemented with a law of conservation of energy or entropy

(37) $$\partial_\alpha\, h^\alpha(u) = g(u)$$

that must be compatible with (3) and therefore [83]

(38) $$\nabla h^i = \nabla h^0 A^i.$$

This is the starting point of the fundamental paper of Friedrichs and Lax. Differentiating in the direction of an arbitrary vector v and multiplying by w one obtains

$$\check{v}\nabla\nabla h^i w - \nabla h^0 \nabla\nabla f^i v w = \check{v} H A^i w,\quad H := \nabla\check{\nabla} h^0.$$

Exchanging v and w does not change the first member. Hence $\overline{A}^i := H A^i$ are symmetric matrices and the system is symmetrized by multiplication

(39) $$H u_t + \overline{A}^i u_i = H f.$$

One can only wonder at the fact that long-established theories of natural philosophy [84], [85] possess this so propitious mathematical structure.

Let us introduce the field

(40) $$u' = \check{\nabla} h^0.$$

By (39) it satisfies the equation

$$u'_t + \check{A}^i u'_i = H f$$

which is not specially interesting since it is neither symmetric nor conservative. We multiply it by H^{-1} that is we go back to the original system making the Le Gendre [91] transformation

(41) $$h'^0 = \check{u}'u - h^0$$

and defining

(42) $$h'^i = \check{u}'f^i - h^i$$

so that, by (38), (40),

$$dh'^\alpha = \check{f}^\alpha du'.$$

As a result [92], [93], ([22], p.42)

$$f^\alpha = \check{\nabla}' h'^\alpha$$

and (3) becomes

(43) $$\partial_\alpha \check{\nabla}' h'^\alpha = f, \quad A'^\alpha u'_\alpha = f, \quad A'^\alpha = \nabla' \check{\nabla}' h'^\alpha$$

that is the symmetric conservative system found by Godunov. Multiplying (43) by u' he deduced (37) where the h^α are given by (41), (42). We shall see later that it is convenient to consider the components of u', called the *main field* by Ruggeri and Strumia [92], as Lagrange multipliers [83], [94].

Now, according to the result of Friedrichs and Lax wherever there is an additional law we may look for the main field u' and the generating functions h'^α. Also for complex fields [95].

A) Relativistic fluid

It is even simpler than the classical case. In a covariant formalism [92] it is easy to check that

$$Td(rSu^\alpha) = u_\beta dT^{\alpha\beta} - (G+1)\, d(ru^\alpha)$$

and therefore, since

$$dh^\alpha = \check{u}' df^\alpha$$

(44) $$h^\alpha = -rSu^\alpha, \quad \check{f}^\alpha = (ru^\alpha, T^{\alpha\beta})$$
$$h'^\alpha = pu^\alpha/T, \quad \check{u}' = (G+1, -u_\beta)/T.$$

Convexity of h^0 is warranted provided that the quadratic form also considered by Friedrichs [101]

(45) $$Q = \delta\check{u}'\, \delta f^\alpha \xi_\alpha$$

is positive definite for some time-like vector ξ_α. This means here

$$C_P > 0, \quad 0 < (\partial p/\partial\rho)_S \leq 1.$$

The important role played by the inverse of the absolute temperature has already been underlined. It is the *coldness* and u_β/T is the *coldness vector* [98], [99]. The components of the main field u' less familiar than the components of the physical field u are interpreted as the observables of the system [100].

B) Hyperelastic material

In Lagrangian coordinates

$$u = (\rho v_i, F_{ik}, \frac{1}{2}\rho \vec{v}^2 + e), \quad f^j = (T_{ij}, v_i\delta^j_k, v_i T_{ij})$$

where T_{ij} is the first Piola-Kirchhoff stress tensor, F_{ij} the displacement gradient tensor, e the internal energy, ρ the constant mass density of the reference configuration and

$$de = TdS - T_{ij}dF_{ij}.$$

The entropy law is simply

$$\partial_t S = 0$$

so that $h^0 = -S, h^i = 0$. It follows

$$dh^0 = -dS = -\frac{1}{T} \left\{ d(\frac{1}{2}\, \rho v^2 + e) - \rho \vec{v}.d\vec{v} + T_{ij} dF_{ij} \right\}$$

and

$$u' = \frac{1}{T}\, (v_i, -T_{ik}, -1),\ h'^0 = \frac{1}{T}\, (\frac{1}{2}\, \rho v^2 - T_{ij} F_{ij} - e + TS).$$

Furthermore, it is easily seen that h^0 is convex with e and that the generators (42) are given by [96], [97]

$$h'^j = \frac{1}{T}\, T_{ij} v_i.$$

C) *Euler variational equations*

With a Lagrangian depending on K functions $q^s(x^\alpha)$ and their first order derivatives $q_\alpha^s := \partial_\alpha q^s$ the Euler-Lagrange equations

(46) $$\partial_\alpha(\partial L/\partial q_\alpha^s) - \partial L/\partial q^s = 0$$

can be written as a first order system (3) where

(47)
$$\check{u} = (\partial L/\partial q_0^s, q_i^s, q^s),$$
$$\check{f}^j = (\partial L/\partial q_j^s, -q_0^s \delta_i^j, 0),$$
$$\check{f} = (\partial L/\partial q^s, 0, q_0^s).$$

A consequence of (46) is the conservation of the energy-momentum tensor

$$\partial_\alpha T_\beta^\alpha = 0,\ T_\beta^\alpha = q_\beta^r\, \partial L/\partial q_\alpha^r - L\delta_\beta^\alpha.$$

The components T_i^0 cannot be convex functions of u. Therefore, we take the density of energy T_0^0 for h^0 and $T_0^i = h^i$

$$dh^0 = \partial L/\partial q_0^s\, dq_0^s + q_0^s\, d(\partial L/\partial q_0^s) - dL = q_0^s\, d(\partial L/\partial q_0^s) - \partial L/\partial q_i^s\, dq_i^s - \partial L/\partial q^s\, dq^s.$$

It follows that

$$u' = (q_0^s, -\partial L/\partial q_i^s, -\partial L/\partial q^s)$$

and comparing with (47) the system is written in the form

(48) $$H'(u')\, u_t' + A'^j u_j' = B'u'$$

where the nonlinearity is only to be found in the coefficient $H' = \nabla'\check{\nabla}'h'^0$ of the temporal derivative, the constant matrices A'^j, B' are respectively symmetric and skew-symmetric. If L does not depend explicitly on the q's the last components of u can be dropped and the second member of (48) vanishes [82]. With the multiplication rules

$$f^j = A'^j u',\ f = B'u'$$

it is easy to check that [102]

$$B'A'^jB' = 0, \ A'^iA'^jB' + B' \ A'^jA'^i = B'\delta^{ij}$$

and

$$A'^iA'^kA'^j + A'^jA'^kA'^i = \delta^{jk} \ A'^i + \delta^{ik} \ A'^j.$$

This last identity is nothing but the Duffin-Kemmer-Petiau relation first established in the thirties for the Klein-Gordon equation [103]-[107]. Each matrix A'^i is its own (Moore-Penrose) inverse.

D) Principal subsystems.

If we suppose that a component of u', say u'^k, is known, i.e., if we assign some value to it and if we remove from the system the corresponding equation [119]

$$\partial_t u^k + \dots = 0, \ u^k = \partial h'^0/\partial u'^k$$

the remaining ones form a system which has the same symmetric hyperbolic structure (43) (in agreement with the interpretation of u' as the observable quantities of the system). Its characteristic velocities are not larger than those of the original system. This interesting fact has been studied by Tai Ping Liu, Chen and Levermore in connection with dissipation [120], [121]. If the number of equations is increased as in Extended Thermodynamics the largest velocity of wave propagation cannot decrease. In fact it does increase as was shown numerically by Weiss for a large number of moments (up to 15180) [122], [100], or theoretically by Cercignani and Majorana by the study of the Boltzmann equation [123]. The largest speed grows without bounds in the classical case or tends to the light speed in the relativistic case.

In addition to the various subsystems already presented [119], let us consider the one obtained from a relativistic fluid by assigning the last component u_0/T of u' in (44) and by leaving aside the energy equation. Writing

$$u_0/T = 1/T_0$$

it appears that the given function T_0 is the temperature in the rest frame. In the remaining equations

$$\partial_\alpha \ (ru^\alpha) = 0, \ \partial_\alpha T^{\alpha i} = 0, \ i = 1, 2, 3$$

the temperature is replaced by

$$T = T_0/\sqrt{1 - v^2/c^2}.$$

This is the transformation law proposed in 1963, 65 by Ott [124] and Arzeliès [125]. It differs from the formula (in which the square root was a multiplying factor) advanced in 1907 by Einstein and Planck and reproduced in numerous textbooks. In the second edition (1972) of his classic Theory of Relativity [126], Møller adopts the new expression and remarks (p.233) :

«The papers by Ott and Arzeliès gave rise to many controversial discussions in the literature and at the present there is no generally accepted description of relativistic thermodynamics».

6. Constraints

Sometimes a system of N equations

$$\text{(49)} \qquad \partial_t u + \partial_i f^i(u) = f(u)$$

is accompanied by M constraints

$$\text{(50)} \qquad \sigma := \partial_i c^i(u) - c(u) = 0$$

that is equations which do not contain time derivatives. To be compatible with the equations of evolution they must be involutive i.e., there must exist $M \times M$ matrices M^i and R such that [108]

$$\text{(51)} \qquad \nabla c^i A^j + \nabla c^j A^i = M^i \nabla c^j + M^j \nabla c^i$$

$$\text{(52)} \qquad \partial_t \sigma + \partial_i (M^i \sigma) = R\sigma$$

and

$$\text{(53)} \qquad \partial_\alpha \psi^{\alpha\beta} = \varphi^\beta, \ \psi^{\alpha\beta} = -\psi^{\beta\alpha},$$

with

$$\psi^{ij} = \nabla c^i f^j - M^i c^j, \ \psi^{io} = c^i, \ \varphi^o = c, \ \varphi^i = M^i c - \nabla c^i f.$$

For the Einstein equations of general relativity Lichnerowicz [109], [110] has shown that four constraints exist which obey the law (52), with variable matrices. However, when all the shocks are not characteristic it turns out that ∇c^i and M^i must be constant matrices to ensure the compatibility of the Rankine-Hugoniot equations with the constraints. The M^i have no physical meaning, their value can be changed. Usually they are equal to zero and this is the reason why Dafermos assumes [111]

$$\nabla c^i f^j + \nabla c^j f^i = 0.$$

Systems with constraints and additional law of entropy or energy can be symmetrized [112]. The supplementary law is obtained by multiplying (49) by $\check{u}' = \nabla h^0$ and (50) by $b(u')$ so that

$$\text{(54)} \qquad \nabla h^i = \check{u}' A^i + b \nabla c^i, \ bc + \check{u}' f = g.$$

As in the preceding paragraph, according to Friedrichs and Lax it appears that the matrices

$$\overline{A}^i = H A^i + \check{\nabla} b \nabla c^i$$

are symmetric and therefore by simple multiplication (49) can be rewritten

$$H u_t + \overline{A}^i u_i = H f + \check{\nabla} bc.$$

Alternatively the change of variables $u \to u'$, the Le Gendre transformation (41) and the introduction of the generating functions

$$h'^i = u' f^i + bc^i - h^i$$

leads to

$$u = \check{\nabla}' h'^0, \ f^i = \check{\nabla}' h'^i - \check{c}^i \check{\nabla}' \check{b}$$

and by substitution in (49), taking into account (50)

(55) $\qquad H'u'_t + A'^i u'_i = f + \check{c}\check{\nabla}'\check{b}, \ H' = \nabla'\check{\nabla}'h'^0, \ A'^i = \nabla'\check{\nabla}'h'^i - \check{c}^i\nabla'\check{\nabla}'\check{b}$

showing that the A'^i's are linear combinations of Hessian matrices.

If (49) is multiplied by ku' (k constant) h^α will be changed in kh^α provided that b is replaced by kb. Thus

$$b(ku') = kb(u')$$

i.e., b is a homogeneous function of u' and

$$\nabla'bu' = b.$$

As a consequence we still have

$$h^\alpha = \nabla'h'^\alpha u' - h'^\alpha.$$

This is the additional law given by Godunov for the symmetric system (55) he obtained for magnetohydrodynamics by direct calculation [113].

Differentiating f^i with respect to u' we obtain the relation between A^i and A'^i

(56) $\qquad\qquad\qquad\qquad A^i H' = A'^i - \check{\nabla}'b\nabla'c^i.$

It follows that while (55) is certainly hyperbolic (at least as long as convexity of h^0 holds) the same is not necessarily true of (49) [114].

As can be seen, for instance, for the well-known Lundquist equations [115] ; the matrix A_n lacks one eigenvector corresponding to $\lambda = 0$ when the velocity of the fluid and the magnetic field are not both zero. It lacks two in the case of Landau's superfluid equations [116], [117]when the direction of propagation $\vec{n}$ is orthogonal to the velocity of the normal part of the fluid but not to the superfluid velocity. *Thus the symmetrization also hyperbolizes these systems by integrating the constraints* [118].

To the constraint of the superfluid

$$\text{curl } \vec{v}_s = 0, \ c^i = \{\vec{e}^i \times \vec{v}_s\}$$

is associated a three-dimensional vector b

$$b = \{(\rho_s/T)\, \vec{v}_n \times \vec{v}_s\}.$$

It follows that A'^i contains four Hessian matrices. In another version of the theory, due to Putterman [127] another curl vanishes and one could expect A'^i to contain seven Hessian matrices but it turns out that the constraints are not effective, i.e., that they need not be considered to deduce the entropy law, in one word that $b = 0$, the simplest structure [118]. Non effective constraints are also those

$$\partial_i q^s_j - \partial_j q^s_i = 0, \ \partial_i q^s - q^s_i = 0$$

of Euler's variational equations as already seen in §5c.

On the other hand if the nonlinear system

$$\partial_t v + f(\partial_j v, v) = 0$$

is not already symmetric [128] its eventual symmetrization, by means of an additional law may involve a large number of Hessian matrices [129].

By (51), (56) we have

(57) $$\nabla c^i A'^j + \nabla c^j A'^i = L^i \nabla' c^j + L^j \nabla' c^i, \quad L^i = M^i + \nabla c^i \check{\nabla}' b.$$

If K is a constant $N \times M$ matrix we can make the change

$$f^i \longrightarrow f^i + Kc^i, \ f \longrightarrow f + Kc$$

then

(58) $$A^i \longrightarrow A^i + K\nabla c^i, \ M^i \longrightarrow M^i + \nabla c^i K$$

and the additional law (37) is left unchanged provided that

$$\check{u}' \longrightarrow \check{u}', \ b \longrightarrow b - \check{u}'K.$$

Hence L^i is also unchanged : $L^i \longrightarrow L^i$.

Assume now that s is not an eigenvalue of L_n. Then it is not difficult, with the help of (56), (57) to check that [130]

(59) $$\nabla' f_n - s\nabla' u = (A_n - sI) \ H' = Q(A'_n - sH')$$

where

$$Q = I - \check{\nabla}' b(L_n - sI)^{-1} \nabla c_n$$

has an inverse

$$Q^{-1} = I + \check{\nabla}' b(M_n - sI)^{-1} \nabla c_n$$

provided that s is not an eigenvalue of M_n a not very stringent condition in view of (58).

Further if λ is a physically meaningful velocity i.e., if the disturbances moving at this speed are compatible with the constraints

$$\delta c_n = 0 \Longrightarrow \nabla c_n d_I = 0, \ I = 1, 2, ..., m$$

then by (59) the eigenvectors of A_n and A'_n/H' are related by

$$d_I = H'd'_I, \ l'_I = l_I Q$$

and it is a matter of straightforward calculation to show that

$$l_I A^\alpha d_{I'} = l'_I A'^\alpha d'_{I'}.$$

For $\alpha = 0$ this relation allows for the choice of an orthonormal system of eigenvectors

$$l'_I = \check{d}'_I, \ l_I d_{I'} = \check{d}'_I H' d'_{I'} = \delta_{II'}.$$

7. Shocks

When, even starting with continuous data, the solution after a while becomes discontinuous the differential equations cease to be valid and the system (3) is replaced by the Rankine-Hugoniot relations [132]

(60) $$f_n(u) - su = f_n(u_0) - su_0, \ f_n := f^i n_i$$

which link the values u and u_0 (the known state before the shock) of the field on both sides of the shock front at a point where its normal velocity is $s\vec{n}$. Additional conditions are added for the constraints

$$c_n(u) = c_n(u_0).$$

There is an abundant literature on the difficult problem of solving (60) (see for instance [2]-[6], [24], [133]-[135]). Here we will restrict ourselves to the use of the main field.

Denoting the jump of any function $q(u)$ by

$$[q] := q(u) - q(u_0)$$

a shock is present if and only if $[u] \neq 0$, or equivalently $[u'] \neq 0$.

A) Global invertibility.

Indeed there is a one to one correspondence between u and u' as is easily seen from the following theorem [136] :

There is a one to one correspondence between x and $y = f(x)$ if the symmetric part of the matrix $\partial_x f$ is positive (or negative) definite : $\delta \check{x} \delta y > 0(< 0)$.

Set

$$x = x_0 + t(x_1 - x_0), \quad \phi(t) = (\check{x}_1 - \check{x}_0)\{f(x) - f(x_0)\}.$$

Then

$$(d/dt)\,\phi(t) = (\check{x}_1 - \check{x}_0)\,\partial_x f(x_1 - x_0) > 0. \quad x_1 \neq x_0$$

and since $\phi(0) = 0$, $\phi(1) = (\check{x}_1 - \check{x}_0)\{f(x_1) - f(x_0)\} > 0$. Hence, $f(x_1) \neq f(x_0)$ if $x_1 \neq x_0$.

Consider the following linear change of variables

$$\begin{aligned} y_1 &= -x_2 \\ y_2 &= x_1 \end{aligned} \implies \partial_x f = \begin{vmatrix} 0 & -1 \\ 1 & 0 \end{vmatrix}$$

the symmetric part is 0 and the theorem does not apply. However, if we multiply the first equation by -1 and commute it with the second we obtain the new functions : $f_1 = x_1$, $f_2 = x_2$. The symmetric part is the unit matrix and the result applies.

In the change $u \to u'$ the quadratic form $\delta \check{u}' \, \delta u$ is invariant under the mentioned transformations and expresses the convexity of h^0. More generally the quantity (45) does not change either : if one equation of the system is multiplied by k the corresponding component of u' is divided by k and if two equations are exchanged so are also their Lagrange multipliers.

Then another question arises : can we consider any of the $N + 1$ conservation laws (3), (37) as the supplementary law? To see it let us pick an equation out of (3) and split the system

$$(S)\ \partial_\alpha f^\alpha = 0 : \begin{cases} \partial_\alpha \, \overline{f}^\alpha &= 0 \\ \partial_\alpha \, g^\alpha &= 0 \end{cases}$$

The additional law

(61)
$$\partial_\alpha h^\alpha \equiv \overline{u}' \, \partial_\alpha \, \overline{f}^\alpha + v' \, \partial_\alpha \, g^\alpha = 0$$

is obtained by multiplying (S) by $\check{u}' = (\overline{u}', v')$. Convexity of h is expressed by

$$Q = \xi_\alpha \, (\delta \overline{u}' \, \delta \overline{f}^\alpha + \delta v' \, \delta g^\alpha) > 0.$$

Now let us consider the system

$$(S_*) \quad \begin{cases} \partial_\alpha \, \overline{f}^\alpha &= 0 \\ \partial_\alpha \, h^\alpha &= 0 \end{cases}$$

with the additional law

$$\partial_\alpha \, g^\alpha = 0$$

obtained, in virtue of (61), by multiplying (S_*) by $\check{u}'_* = (-\overline{u}', 1)/v'$.

Convexity of g implies

$$Q_* = \xi_\alpha \, \{ -\delta \, (\overline{u}'/v') \, \delta \overline{f}^\alpha + \delta \, (1/v') \, \delta h^\alpha \} = -Q/v' > 0$$

that is $v' < 0$. If $v' > 0$, g is concave but $-g$ is convex. Therefore, the condition, already expressed by Friedrichs and Lax [83], to exchange the additional law is $v' \neq 0$. This requirement can also easily be met if v' is bounded : replacing h^α by $h^\alpha + k g^\alpha$ in (61) changes v' in $v' + k$. For a fluid the law of entropy and energy can therefore be exchanged since $v' = -1/T < 0$ (36). However, (see section C below) as soon as weak solutions are concerned the choice of the additional law is not indifferent since it does not satisfy the Rankine-Hugoniot jump condition. While it would seem that the conservation of entropy would belong to (S) the relativistic formulation of the fluid clearly shows that this equation stands apart.

The generating vectors $h'^\alpha(u')$ of (S) and $h_*'^\alpha(u_*')$ of (S_*) are simply related

$$h_*'^\alpha = -(\overline{u}'/v') \, \overline{f}^\alpha + (1/v') \, h^\alpha - g^\alpha = -h'^\alpha/v'.$$

Related to the existence of additional laws is the restriction of solutions by means of an *equation of state* : is it possible to assign a relation $\phi(u) = 0$ compatible with the field equations? If it is the derivative

$$\partial_t \phi = \nabla \phi u_t = -\nabla \phi A^i u_i$$

must vanish when

$$\partial_i \phi = \nabla \phi u_i$$

does (involution). This implies the existence of a vector $\Lambda^i(u)$ such that

$$\nabla \phi \, (A^i - \Lambda^i \, I) = 0.$$

In general, this will not be possible unless $\nabla \phi$ is a left eigenvector for all the A^i's. Then an additional law $h^0 = \phi$, $h^i = \phi \Lambda^i$ exists

$$\partial_t \phi + \partial_i \, (\phi \Lambda^i) = \phi \partial_i \Lambda^i = 0.$$

In (S) $\partial_\alpha g^\alpha = 0$ may then be considered as the supplementary law of the system $\partial_\alpha \overline{f}^\alpha = 0$ of $N - 1$ equations for the reduced $N - 1$ field variables. Wave velocities are unchanged : $\delta \phi \equiv 0$ for any disturbance moving at a speed different from $\vec{\Lambda} \cdot \vec{n}$. The multiplicity of this speed is reduced by one unity.

For a fluid

$$\phi = -\rho\,(S - S_0), \quad \Lambda^i = u^i$$

and $\phi = 0$ represents the isentropic fluid ($S = $ const.) the equations of which can be symmetrized with the help of the energy equation.

B) Subluminal velocities

Conditions (60) can be read

$$y(u', s) = y(u'_0, s).$$

The symmetric part of $\nabla' y$ is just

$$A'_n - sH'$$

and when s lies outside the range of the eigenvalues (wave velocities) of A'_n/H' the matrix $\nabla' y$ is definite and the only solution of the Rankine-Hugoniot equations is $u = u_0$. It follows that the speed of the shock is limited by the extrema of the wave speeds and since in a relativistic theory these do not exceed the light velocity this is also true for the shocks [137].

In the presence of constraints [130] we introduce the variable

$$\breve{u}''(u') = \breve{u}' + b(u')\,(M_n - sI)^{-1}\nabla c_n = \breve{u}'Q^{-1}$$

assuming $\det\,(M_n - sI)\,\det\,(L_n - sI) \neq 0$. Hence,

$$\breve{\nabla}'\breve{u}'' = Q^{-1} \Longrightarrow \breve{\nabla}''\breve{u}' = Q$$

and by (59)

$$\nabla'' y = \nabla' y \nabla'' u' = Q(A'_n - sH')\breve{Q}.$$

As above if the speed is too large there is no shock in u'' which also means no shock in u satisfying the constraints

(62) $$\nabla c_n(u - u_0) = 0$$

for

$$(\breve{u}'' - \breve{u}''_0)\,(u - u_0) = (\breve{u}' - \breve{u}'_0)\,(u - u_0) > 0, \quad (u''_0 = u''(u'_0))$$

whenever $[u] \neq 0$ (because of convexity).

In practice the following criterion [138] is of convenient use for relativistic theories

Suppose that for any time-like vector

$$Q = \xi_\alpha \delta\breve{u}'\,\delta f^\alpha > 0, \quad \xi_\alpha \xi^\alpha > 0$$

for any $\delta u'$ compatible with the constraints

$$\xi_i \delta c^i = 0$$

and that

$$\det\,(L^\alpha \xi_\alpha) > 0, \quad \det\,(M^\alpha \xi_\alpha) > 0\;(L^0 = M^0 = I)$$

then the wave and shock velocities do not exceed the velocity of light.

Convexity of $h^0(u)$ (hence hyperbolicity) is included as a special choice ($\xi_0 = 1, \xi_i = 0$). An application to magnetohydrodynamics follows in §8.

C) Generating function for the shocks

For linear equations the solution of (60) is simply

$$u = u_0 + u^I d_I, \quad s = \lambda.$$

In general s will not be constant and we may assume it depends also on the parameters u^I.

If we insert this solution $u = u(u^I, u_0), s = s(u^I, u_0)$ in the jump equation of the additional law (37) we obtain a function

$$\eta(u^I, u_0) = [h_n] - s[h^0].$$

For an ordinary shock depending on one parameter, η which vanishes with the shock is otherwise positive. Dafermos recalls that this entropy condition was first introduced, in gas dynamics, by Jouguet. It is associated with the admissibility conditions of Lax [26], [134],

$$(63) \qquad \lambda^k(u_0) < s < \lambda^{k+1}(u_0), \quad \lambda^{k-1}(u) < s < \lambda^k(u)$$

for non characteristic shocks and strictly hyperbolic systems. This is not always the case [139], [152]. See the papers by Dafermos [140] and Liu [141] on this important question. η generates the shock.

On one hand as a function of s and $y_0 = f_n(u_0) - su_0$ it satisfies a Hamilton-Jacobi equation and the shock solutions are singular integrals of this equation [142].

On the other hand previous results are easily generalized to the case with constraints [143]. We have, thanks to the Friedrichs-Lax condition (38), (54),

$$(64) \qquad \partial_I \eta = w \partial_I s, \quad \partial_I w = [\check{u}] \, \partial_I u'$$

$$(65) \qquad \nabla_0 \eta = [\check{u}''] \, (A_{on} - sI) + w \nabla_0 s, \quad A_{on} = A_n(u_0)$$

where

$$w := \check{u}' \, (u - u_0) - h^0(u) + h^0(u_0)$$

is positive and vanishes only with the shock. When the shock tends to zero the u^I's tend to zero and the velocity s tends to some eigenvalue $\lambda_0 = \lambda(u_0)$. (The converse is not always true [144]).

The quantity

$$\hat{\lambda}(u_0, u_1) = [\check{u}''] \, [f_n]/[\check{u}''] \, [u]$$

is physically meaningful ; the terms of each scalar product have the same dimension. The ratio is a velocity. Then

$$[\check{u}''] \, [u] \, \nabla_0 \hat{\lambda} = ([f_n] - \hat{\lambda}[u]) \, \nabla_0 u_0'' + [\check{u}''] \, (A_{0n} - \hat{\lambda}I).$$

For a solution of the Rankine-Hugoniot equations $\hat{\lambda}$ is the velocity of the shock and

$$[\breve{u}''] \, (A_{0n} - \hat{\lambda}I) = -[\breve{u}''] \, [u] \, \nabla_0 \hat{\lambda}$$

a formula also valid for the mean eigenvalue $\tilde{\lambda}(u_0, u_1)$ [145]. The second order derivatives of $\hat{\lambda}$ and $\tilde{\lambda}$ are different.

D) Ordinary shocks

When $\partial_I s \neq 0$ (64) shows that η depends on the sole parameter s and starting from a null shock we have by means of the Lax conditions (63)

$$\partial_s \eta = w, \quad \eta = \int_{\lambda_0}^{s} w \, ds > 0$$

a result that can also be obtained by artificial viscosity [146], [147] and which is known as the increase of entropy. In fact, for a fluid,

$$\eta = \rho_0(s - u_{on}) \, [S] > 0 \Longrightarrow [S] > 0$$

for a relative positive velocity. The function η has several branches each one with an inflexion point at some non-degenerate velocity λ_0. In this neighbourhood [26]

$$\eta = \alpha(s - \lambda_0)^3 + ...$$

as it is easily seen from (64) that

$$\partial_{ss}\eta = \partial_s w = (\breve{u} - \breve{u}_0) \, \partial_s u'$$

vanishes with the shock. The function η has been studied for several theories [148]-[152], ([100], pp.123, 157, 166). Its graph allows for a representation of the various shocks originating from various eigenvalues and its slope measures in some way the strength of the shock.

Now since s is taken as the parameter u^1 (65) reduces to

$$[\breve{u}''] \, (A_{on} - sI) = \nabla_0 \eta \, (s, u_0)$$

yielding the jump of u''. For small shocks

$$[\breve{u}''] = (s - \lambda_0) \, v_1 + (s - \lambda_0)^2 v_2 + ..., \nabla_0 \eta = -3\alpha \, (s - \lambda_0)^2 \nabla_0 \lambda_0 + ...$$

and substituting in the preceding equation we obtain

(66)
$$[\breve{u}''] = \omega \ell_0 + w g_0 + ...$$

up to terms of third order, where $g(u)$ is defined by

$$g \, (A_n - \lambda I) = -\nabla\lambda + (\nabla\lambda d) \, \ell, \quad g d = 0, \quad \ell d = 1,$$

or [155]

(67)
$$g = \sum_i{}' \, (\nabla\lambda \cdot d^i) \, \ell^i / (\lambda - \lambda^i), \quad \ell^i d^j = \delta^{ij}$$

with a summation over all the eigenvectors not corresponding to λ.

E) Characteristic shocks.

In this case $\partial_I s = 0, \partial_I \eta = 0, s$ and η have the values they take when the shock is null i.e.,

$$s = \lambda_0, \quad \eta = 0$$

and from (65)

(68) $$[\breve{u}''] (A_{on} - \lambda_0 I) = -w \nabla_0 \lambda_0$$

which implies

$$\nabla \lambda d_I = 0$$

for $u = u_0$, i.e., λ is exceptional. By (67), (68)

(69) $$[\breve{u}''] = \omega^I \ell_{I^0} + w g_0, \quad \omega^I = \omega^I(u^{I'}, u_0).$$

which is the explicit solution of the characteristic shock. The structure is that of a weak shock (66). To study w as a function of the parameters u^I we consider in a $m + 1$ dimensional space the surface $f(u^I, w) = 0$ where

$$f(u^I, w) := \breve{u}'(u - u_0) - h^0(u) + h^0(u_0) - w$$

and compute the derivatives considering w and the u^I as independent variables and taking into account (62), (69)

$$\partial f / \partial u^I = \partial \breve{u}' / \partial u^I [u] = \omega_I^{I'} \ell_{I'^0}[u], \quad \omega_I^{I'} = \partial \omega^{I'} / \partial u^I$$

$$\partial f / \partial w = \partial \breve{u}' / \partial w [u] - 1 = g_0[u] - 1,$$

$$\partial^2 f / \partial u^I \partial u'^I = \omega_{II'}^{I''} \ell_{I''^0}[u] + \partial \breve{u}' / \partial u^I H' \partial u' / \partial \breve{u}^{I'},$$

$$\partial^2 f / \partial u^I \partial w = \partial \breve{u}' / \partial u^I H' \partial u' / \partial w,$$

$$\partial^2 f / \partial w^2 = \partial \breve{u}' / \partial w H' \partial u' / \partial w.$$

If we commute u and u_0 in w we obtain a quantity w' which is positive

$$w' = [\breve{u}'] [u] - w = \omega^I \ell_{I^0}[u] + w(g_0[u] - 1) > 0, \quad [u] \neq 0.$$

If the shock depends effectively on m parameters, $\det \omega_I^{I'} \neq 0$ and this inequality shows that all points of the surface $f = 0$ are regular points. If the equations

(70) $$\ell_{I^0}[u] = 0, \quad g_0[u] - 1 > 0$$

have a solution w has an absolute maximum (the matrix of the second order derivatives is positive definite at this point) and the shock is bounded (Alfvén shock for instance) [143]. In this case a non-characteristic shock may start from $s = \lambda_0$ with this maximum value [144]. This happens for a nonlinear dielectric medium [152]. See also the von Kármán-Tsien fluid [153] and magnetohydrodynamics [154].

For the variational equations (§5c) there is a linear link between the jumps of u and u'. From the equations

$$\partial_t u + A'^i u'_i = B' u', \quad \partial_t h^0 + \partial_i (\frac{1}{2} \breve{u}' A'^i u') = 0$$

we deduce

$$\lambda_0[u] = A'_n \,[u'], \ \ \lambda_0[h^0] = \frac{1}{2}\,[\check{u}'A'_n u'],$$

$$2\lambda_0 w = 2\check{u}'A'_n \,[u'] - [\check{u}'A'_n u'] = [\check{u}']\,A'_n \,[\check{u}']$$

and by substitution of (69) (now $b = 0$),

(71) $$\alpha_0 w^2 - 2w + |u|^2 = 0$$

where

$$\alpha_0 = \alpha(u_0) = g_0 A'_n \,\check{g}_0/\lambda_0, \ \ |u|^2 = \sum_I (u^I)^2$$

and the w^I's have been taken as parameters : $w^I = u^I$.

If $\alpha_0 > 0$ (71) is an ellipse and the shock is bounded. Conditions (70) $u^I = 0, \alpha_0 w > 1$ correspond to the maximum $2/\alpha_0$ of w.

In any case the evolution of the shock along the rays can be reduced to linear differential equations [131]. In nonlinear optics the law of rotation of the electromagnetic field after the characteristic shock have been earlier determined by direct calculations [156].

A right eigenvector of A'_n/H' is $\partial_I u'$. Therefore

$$\partial_I \check{u}'A'_n \partial_{I'} u' = \lambda \partial_I \check{u}'H'\partial_{I'}u' = (\ell_{I^0} + \partial_{I}w\,g_0)\,A'_n\,(d'_{I'0} + \partial_{I'}w\,\check{g}_0)$$

and the eigenvectors

$$\ell'_I = (\delta_I^{I'} - u_I\,u^{I'}/w)\,(\ell_{I'0} + \partial_{I'}w\,g_0), \ \ \ u_I = u^I$$

are orthonormal

$$\ell'_I\,H'\,d'_{I'} = \delta_{II'}$$

provided they are when $u = u_0$.

F) Crossing eigenvalues

When multiplicity is due to the crossing of two or more eigenvalues λ^1, λ^2 for some value $\vec{n} = \vec{n}_0(u)$ depending on the field. Here the important criterion is the exceptionality of the difference of the velocities [35]

$$\nabla(\lambda^1 - \lambda^2)\,d_I = 0, \ \ \vec{n} = \vec{n}_0(u).$$

Hence, a relative term completes the expression (69) [157]

(72) $$[\check{u}''] = u^I \ell_{I^0} + w g_0 + \sigma g_0^\tau$$

where

$$g^\tau(A_n - \lambda I) = -\nabla(\lambda^1 - \lambda^2), \ \ g^\tau d_I = 0.$$

The shock is bounded if in addition to (70) the jump of u satisfies

$$g_0^\tau[u] = 0.$$

For the variational equations the ellipse (71) is then replaced by an ellipsoid because of the additional variable σ.

In magnetohydrodynamics [1], [2], [158], [159] no less than five velocities coalesce to the value $\vec{v} \cdot \vec{n}$ when the magnetic field is tangent to the wave front. The various vectors of formula (72) have been computed in this case [160].

G) Strictly exceptional systems

If all the velocities are linearly degenerated the system is completely exceptional. If, moreover, it admits (without any selection criterion) no shock but characteristic shocks we call it strictly exceptional [161].

Of special interest are the systems

$$\partial_t w^i + \lambda^i(w)\, \partial_x w^i = 0, \quad i = 1, 2, ..., N.$$

They have been extensively studied by Denis Serre [7] under the name of rich systems because of their many properties. They are the diagonalizable systems [8] and possess an infinity of conservation laws. As a special case a completely exceptional system of two equations can be rewritten in the above form [162] in terms of the velocities λ, μ

$$\partial_t \lambda + \mu \partial_x \lambda = 0$$
$$\partial_t \mu + \lambda \partial_x \mu = 0$$

or in conservative form

$$\partial_t f(\lambda, \mu) + \partial_x g(\lambda, \mu) = 0$$
$$f = \frac{M(\mu) - L(\lambda)}{\lambda - \mu}, \quad g = \frac{\lambda M(\mu) - \mu L(\lambda)}{\lambda - \mu},$$

with the Rankine-Hugoniot conditions

(73) $$[g] - s[f] = 0$$

and the Lax conditions

(74) $$\lambda_0 < s < \mu_0, \quad s < \lambda < \mu$$

for possible non-characteristic shocks depending on the choice of L, M [161].

If we take $(L = 1, M = 0)$, $(L = \lambda, M = 0)$ the two equations (73) only admit characteristic shock solutions, i.e., either $s = \lambda = \lambda_0$, $[\mu] \neq 0$ or $s = \mu = \mu_0, [\lambda] \neq 0$. Such a system is strictly exceptional.

If the second equation is replaced by $(L = -\lambda^2, M = -\mu^2)$ besides the characteristic shocks exists the solution

$$s = \frac{1}{2}(\lambda + \lambda_0) = \frac{1}{2}(\mu + \mu_0), \quad [\lambda] \neq 0, \ [\mu] \neq 0$$

which does not satisfy (74).

Finally, with $(L = 2\lambda - 3, M = 0)$, $(L = 2\lambda^2 - 3, M = 0)$ the following non-characteristic shock

$$s = \frac{1}{2}, \quad \lambda_0 = 0, \quad \mu_0 = 1, \quad \lambda = 1, \quad \mu = 2$$

satisfy both (73) and (74). Furthermore, since the general solution

$$\lambda = 3(1 - \lambda_0)/(3 - 2\lambda_0)$$

is independent of s this kind of shock cannot vanish for some value of s. And this might be a requirement to make.

To the class of strictly exceptional systems belong, besides, of course, the linear systems, the Einstein equations for vacuum, Born-Infeld nonlinear electrodynamics [163], the von Kármán-Tsien fluid when a and b in (31) are constant [165], [166] but not when they depend on S [153], the incompressible relativistic fluid [88], [90], the relativistic string [164].

H) Interaction with a wave

The problem is to find the amplitudes of the reflected and transmitted weak discontinuities when an incident wave impinges on a shock. Here the Lax conditions (63) play also an essential role. It appears that such a perturbation creates a discontinuity in the acceleration of the shock front [167], [168]. For a characteristic shock this acceleration is not an unknown yet the amplitudes are completely determined [152], [169], [170].

8. Relativistic magnetohydrodynamics

The mathematical structure (including the main field u' and the potential vector h'^{α}) of relativistic magnetohydrodynamics has been determined by Anile and Pennisi [172], [173]. The field equations express the null divergence of a vector V^{α} (conservation of mass), and of two tensors, a symmetric one $T^{\alpha\beta}$ (energy-momentum), the other one skew-symmetric $\psi^{\alpha\beta}$ (Maxwell equations)

$$(75) \qquad \partial_{\alpha} V^{\alpha} = 0, \; \partial_{\alpha} T^{\alpha\beta} = 0, \; \partial_{\alpha} \psi^{\alpha\beta} = 0, \; \alpha, \beta = 0, 1, 2, 3.$$

For $\beta = 0$ the last equation is a constraint,

$$(76) \qquad \partial_i \, \psi^{io} = 0, \; i = 1, 2, 3$$

corresponding to div $\vec{B} = 0$ in the classical case.

Now thanks to the Gibbs relation

$$di = T dS + V dp$$

$$Td(-rSu^{\alpha}) = (G + 1) \, dV^{\alpha} - u_{\beta} dT^{\alpha\beta} - B_{\beta} d\psi^{\alpha\beta}, \; G = i - TS, \; r = 1/V.$$

It follows with

$$h^{\alpha} = -rSu^{\alpha}, \; f^{\alpha} = (V^{\alpha}, \; T^{\alpha\beta}, \; \psi^{\alpha i})$$

that

$$u' = (G + 1, -u_{\beta}, \; -B_i)/T, \; b = -B_0/T, \; h'^{\alpha} = \left(p + \frac{1}{2} B^2\right) u^{\alpha}/T.$$

The constraint $c^i(u) = \psi^{io}$ is linear in $u := f^0$ and $c^i(f^j) = \psi^{ij}$ so that (§6) $M^i = 0, L^i = u^i/u^0$. According to (§7,b) the quadratic form

$$\mathcal{Q} = \xi_{\alpha} \delta \left(\frac{G + 1}{T}\right) \delta(ru^{\alpha}) - \xi_{\alpha} \delta \left(\frac{u_{\beta}}{T}\right) \delta T^{\alpha\beta} - \xi_{\alpha} \delta \left(\frac{B_{\beta}}{T}\right) \delta \psi^{\alpha\beta}$$

must be positive definite. The simplest is to compute this invariant in the rest frame of the particle. The conditions found [174] are those of the ordinary fluid

$$(77) \qquad\qquad C_p > 0, \ 0 < \left(\frac{\partial p}{\partial \rho}\right)_S \le 1.$$

The subluminal character of waves and shocks are thus ensured. Lichnerowicz [175] had already arrived at this conclusion under the very simple hypothesis

$$\tau_p < 0, \ \tau_{pp} > 0, \ \tau\ (p, S) = (1 + i)\ V$$

which also holds for the fluid [176]. Expressed with $(\partial p / \partial \rho)_S = p'$ these are just [87]

$$\tau_p = V^2(1 - 1/p') < 0, \ \tau_{pp} = V^2\{(\rho + p)\ p'' + 2p'\ (1 - p')\}/(\rho + p)\ p'^3 > 0.$$

The first inequality is the second of (77) while the second means by (32) that the sound wave of the fluid is not linearly degenerated. For details the reader is referred to the new version of Lichnerowicz's book [177]. A generalization of the magnetohydrodynamics equations results from the coupling of the fluid with the Yang-Mills equations [178].

In order to eliminate the constraint (76) Pennisi [179] introduces a new variable X through

$$(78) \qquad\qquad \partial_\alpha\ (\psi^{\alpha\beta} + Xg^{\alpha\beta}) = 0$$

which replaces the last equation (75). He then determines a new potential vector

$$h'^\alpha\ (u') = \{p + \frac{1}{2}\ (B^2 + X^2)\}\ u^\alpha/T - X B^\alpha/T$$

which differentiated with respect to the new main field

$$u' = (G + 1, -u_\beta, \ -B_\beta + Xu_\beta)/T$$

yields in addition to (78) and $\partial_\alpha V^\alpha = 0$ the new field equations

$$\partial_\alpha\ (T^{\alpha\beta} + X\psi^{\alpha\beta} + \frac{1}{2}\ X^2 g^{\alpha\beta}) = 0$$

The conservation of entropy is unchanged

$$\partial_\alpha\ (SV^\alpha) = 0.$$

This system has a simpler structure. The A'^α's are Hessian matrices and the solutions of (75) are obtained by just taking $X = 0$ at some instant. Furthermore, the condition $Q > 0$ is satisfied under the same restrictions (77). Besides subluminal wave and shock velocities are unchanged. Now in virtue of (53) equation (78) can also be written for several involutive constraints with a multicomponent X. A procedure was suggested [180] to find the new main field and the new potential vector. When it works the constraints can be suppressed in this way. It does for magnetohydrodynamics and provides just another derivation of the Pennisi's equations.

9. Generating new equations

In 1948 Cattaneo in a famous paper [181] proposed a new equation to solve the paradox of the infinite velocity of heat propagation. The well-known Fourier equation was then replaced by hyperbolic equations. Slowly the idea made its way and it seemed desirable to have hyperbolic systems compatible with an entropy. Such systems are generated by the h'^α but the knowledge of h^α is not sufficient to determine them [93]. How are they to be found? When a Lagrangian exists it is chosen to be a function of the invariants (through changes of reference frames in space or space-time) of the theory. Now we also need the vectors of the theory. Those which can be constructed with the components of u'. Since they are Lagrange multipliers we know their forms if we know the form of the equations: if a component of f^α is a tensor of order k the corresponding component of u' will be a tensor of order $k - 1$. For instance if the system is to consist of one scalar and one vectorial equation

$$\partial_\alpha V^\alpha = 0, \quad \partial_\alpha T^{\alpha\beta} = 0$$

the u' field will be of the type

$$\breve{u}' = (\omega, v_\beta).$$

Therefore by a theorem of representation h'^α will have the form

$$h'^\alpha = \phi\,(\omega, \nu)\,v^\alpha, \quad \nu = \frac{1}{2}\,v_\alpha v^\alpha$$

and

$$V^\alpha = \partial h'^\alpha/\partial\omega = \phi_\omega v^\alpha$$
$$T^{\alpha\beta} = \partial h'^\alpha/\partial v_\beta = \phi_\nu\,v^\alpha v^\beta + \phi\,g^{\alpha\beta}.$$

The additional law follows

$$\partial_\alpha h^\alpha = 0, \quad h^\alpha = \omega V^\alpha + v_\beta T^{\alpha\beta} - \phi\,v^\alpha = (\omega\phi_\omega + 2\nu\phi_\nu)\,v^\alpha.$$

Such are, for instance, the equations of the relativistic fluid. Although they do not derive from a variational principle (with convex density of energy) they are completely determined by the knowledge of the sole function ϕ.

This method has been first employed by Ruggeri [182] for the rigid conductor (a generalization of Cattaneo's equation) [183]-[185], by Anile, Pennisi and Sammartino [186] for the determination of the Eddington factor of stellar physics and by Müller and Ruggeri for the monoatomic gas out of equilibrium (gas with heat conduction and viscosity). All these results are reported in their book [100]. Of course, the tensorial structure of the field is found on the basis of physical considerations (kinetic mechanics, for instance).

In the next paragraph an application is given to nonlinear electromagnetism.

10. A System of Electromagnetic Type

A) Form of the system

The system

$$\partial_\alpha G^{\alpha\beta} = 0, \quad \partial_\alpha H^{\alpha\beta} = 0, \quad G^{\alpha\beta} = -G^{\beta\alpha}, \quad H^{\alpha\beta} = -H^{\beta\alpha}$$

splits into the evolution equations

$$(79) \qquad \partial_\alpha f^\alpha = 0, \ u = f^0 = \begin{vmatrix} G^{0j} \\ H^{0j} \end{vmatrix}, \quad f^i = \begin{vmatrix} G^{ij} \\ H^{ij} \end{vmatrix}$$

and the constraints

$$(80) \qquad \partial_i c^i = 0, \ c^i = \begin{vmatrix} G^{i0} \\ H^{i0}. \end{vmatrix}$$

To obtain the energy law (79) must be multiplied by the main field $\breve{u}' = (u_j, v_j)$ and (80) by $b = (\mu, \nu)$ so that

$$\breve{u}' \partial_\alpha f^\alpha + b \partial_i c^i = \partial_\alpha h^\alpha.$$

Now with the u' field one can construct the generating vector

$$h'^i = a u^i + b v^i + c \sigma^i, \ \vec{\sigma} = \vec{u} \times \vec{v}, \ u^i = u_i$$

and the generating function

$$h'^0 = w$$

which together with the other scalars $a, b, c, \mu. \nu$ is a function of the three-dimensional invariants

$$p = \frac{1}{2}\vec{u}^2, \ q = \vec{u}.\vec{v}, \ r = \frac{1}{2}\vec{v}^2.$$

It follows that

$$\breve{u} = (G^{0j}, H^{0j}) = \partial h'^0 / \partial u' = (\partial w / \partial u_j, \partial w / \partial v_j)$$
$$= (w_p u^j + w_q v^j, \ w_q u^j + w_r v^j),$$
$$G^{ij} = \partial h'^i / \partial u_j - G^{i0} \partial \mu / \partial u_j - H^{i0} \partial \nu / \partial u_j,$$
$$H^{ij} = \partial h'^i / \partial v_j - G^{i0} \partial \mu / \partial v_j - H^{i0} \partial \nu / \partial v_j.$$

After simple computations the anti-symmetrical conditions

$$G^{ij} + G^{ji} = 0, \ H^{ij} + H^{ji} = 0, \ G^{i0} = -G^{0i}, \ H^{i0} = -H^{0i}$$

imply that, $a = b = 0, c = $ const. while μ, ν disappear. Without restriction one can take $c = 1$ and the field equations become

$$(81) \qquad \partial_t (w_p \vec{u} + w_q \vec{v}) - \text{curl } \vec{v} = 0, \ \partial_t (w_q \vec{u} + w_r \vec{v}) + \text{curl } \vec{u} = 0$$

with the constraints

$$(82) \qquad \text{div } (w_p \vec{u} + w_q \vec{v}) = 0, \ \text{div } (w_q \vec{u} + w_r \vec{v}) = 0.$$

For the energy law

$$\partial_\alpha h^\alpha = 0, \ h^\alpha = \breve{u}' f^\alpha - h'^\alpha$$

which means

$$h^0 = 2(p w_p + q w_q + r w_r) - w, \ h^i = \sigma^i.$$

B) Variational principle

System (81) has the form of the Euler-Lagrange equations (linearity of the f^i). Therefore, we look for a variational principle. Setting

$$\vec{u} = \text{grad } \Phi - \partial_t \vec{A} =: \vec{E}. \quad w_q \vec{u} + w_r \vec{v} = \text{curl } \vec{A} =: \vec{B}$$

to satisfy (81_2), (82_2)

$$(83) \qquad \partial_t \vec{B} + \text{curl } \vec{E} = 0, \quad \text{div } \vec{B} = 0$$

we assume the existence of a Lagrangian $L(\alpha, \beta, \gamma)$

$$\alpha := \frac{1}{2}\vec{E}^2, \quad \beta := \vec{E}.\vec{B}, \quad \gamma := \frac{1}{2}\vec{B}^2$$

and write

$$\partial_\alpha(\partial L/\partial(\partial_\alpha \Phi)) = 0, \quad \partial_\alpha(\partial L/\partial(\partial_\alpha \vec{A})) = 0$$

to obtain

$$(84) \qquad \text{div }(L_\alpha \vec{E} + L_\beta \vec{B}) = 0$$

$$\partial_t(L_\alpha \vec{E} + L_\beta \vec{B}) + \text{curl }(L_\beta \vec{E} + L_\gamma \vec{B}) = 0$$

which are identical to the first equation of (82) and (81) if

$$\vec{v} = L_\beta \vec{E} + L_\gamma \vec{B}, \quad w_p \vec{u} + w_q \vec{v} = -(L_\alpha \vec{E} + L_\beta \vec{B}).$$

Multiplying (83) by $\vec{v}$ and (84) by $-\vec{u}$ the energy law is immediately found with

$$h^0 = L - (2\alpha L_\alpha + \beta L_\beta), \quad h^i = \sigma^i, \quad \vec{\sigma} = L_\gamma \vec{E} \times \vec{B}.$$

Hence

$$w = h'^0 = \breve{u}'u - h^0 = -\vec{E}.(L_\alpha \vec{E} + L_\beta \vec{B}) + \vec{B}(L_\beta \vec{E} + L_\gamma \vec{B}) - h^0 = \beta L_\beta + 2\gamma L_\gamma - L.$$

C) Convexity

If

$$(85)$$
$$Q := \xi_\alpha \delta f^\alpha \delta u' = \xi_0\{\delta \vec{B}\delta(L_\beta \vec{E} + L_\gamma \vec{B}) - \delta \vec{E}\delta(L_\alpha \vec{E} + L_\beta \vec{B})\} + 2\vec{\xi}.\delta \vec{E} \wedge \delta(L_\beta \vec{E} + L_\beta \vec{B})$$

is positive definite for each time-like $\xi_\alpha(\xi_0 > |\vec{\xi}|)$ the system (83), (84) is symmetric hyperbolic and the wave and shock speeds do not exceed the velocity of light.

To get necessary conditions we first choose

$$(86) \qquad \delta\alpha = \vec{E}\delta\vec{E} = 0, \quad \delta\beta = \vec{E}\delta\vec{B} + \vec{B}\delta\vec{E} = 0, \quad \delta\gamma = \vec{B}\delta\vec{B} = 0$$

and (85) reduces to

$$Q/\xi_0 = -L_\alpha \delta\vec{E}^2 + L_\gamma \delta\vec{B}^2 + 2L_\gamma(\vec{v} \times \delta\vec{E}).\delta\vec{B}$$
$$(87) \qquad = L_\gamma(\delta\vec{B} + \vec{v} \times \delta\vec{E})^2 - (L_\gamma + L_\alpha)\,\delta\vec{E}^2 + L_\gamma(1 - \vec{v}^2)\,\delta\vec{E}^2$$
$$+ L_\gamma(\vec{v}.\delta\vec{E})^2 > 0, \quad \vec{v} = \vec{\xi}/\xi_0, \quad \vec{v}^2 < 1.$$

One of the two vectors $\delta\vec{E}, \delta\vec{B}$ is taken to be zero and the other one is parallel to $\vec{E} \times \vec{B}$ yielding the necessary conditions

(88) $$L_\alpha < 0, \quad L_\gamma > 0.$$

The next choice is

$$\delta\vec{B} + \vec{\nu} \wedge \delta\vec{E} = 0, \quad \vec{\nu}.\delta\vec{E} = 0$$

which, by (86) means that $\delta\vec{E}$ must be orthogonal to $\vec{E}, \vec{B} - \vec{E} \times \vec{\nu}, \vec{B} \times \vec{\nu}$. This is the case if we take $\delta\vec{E}$ parallel to $\vec{E} \wedge \vec{\nu}$ and if we define $\vec{\nu}$ by

$$\vec{B} = a\vec{\nu} + \vec{E} \times \vec{\nu}.$$

Then

$$\vec{\nu}^2 < 1 \Longleftrightarrow a^4 + a^2(\vec{E}^2 - \vec{B}^2) - (\vec{E}.\vec{B})^2 > 0$$

while (87) is simply

$$-(L_\alpha + L_\gamma)\, \delta\vec{E}^2 + L_\gamma(1 - \vec{\nu}^2)\, \delta\vec{E}^2 > 0.$$

The second term is small if a^2 is slightly larger than the positive root of the above polynomial. Therefore

(89) $$L_\alpha + L_\gamma \leq 0.$$

Now the general expression (87) can be rewritten with any ϕ

$$Q/\xi_0 = \frac{1}{\phi}\, (\delta\vec{v} - L_\beta\delta\vec{u} + \phi\vec{\nu} \times \delta\vec{u})^2 + \phi(1 - \vec{\nu}^2)\, \delta\vec{u}^2$$

$$+ \phi(\vec{\nu}.\delta\vec{u})^2 + \frac{1}{2\gamma}\, \left(\frac{1}{L_\gamma} - \frac{1}{\phi}\right) \left[(\delta\vec{v} - L_\beta\delta\vec{u}) \wedge \vec{B}\right]^2$$

$$- \frac{1}{2\alpha}\, (L_\alpha + \phi)\, (\delta\vec{u} \wedge \vec{u})^2 + Q'$$

where Q' is a quadratic form in the sole invariants α, β, γ,

(90) $$Q' = -(\delta\alpha\delta L_\alpha + \delta\beta\delta L_\beta + \delta\gamma\delta L_\gamma) - \frac{2}{L_\gamma}\, (\alpha\delta L_\beta^2 + \beta\delta L_\beta\delta L_\gamma + \gamma\delta L_\gamma^2)$$

$$+ \frac{1}{2\gamma}\, (\beta\delta L_\beta + 2\gamma\delta L_\gamma + L_\gamma\delta\gamma)^2\, \left(\frac{1}{L_\gamma} - \frac{1}{\phi}\right) - \frac{1}{2\alpha}\, (L_\alpha + \phi)\, \delta\alpha^2.$$

Assuming (88) and (89) we can choose ϕ such that

(91) $$0 < L_\gamma \leq \phi \leq -L_\alpha$$

then Q is positive definite if, in addition

(92) $$Q' \geq 0.$$

Similar conditions can be obtained for w using the formulation of the first section. Since we can write

$$Q/\xi_0 = \psi\, (\delta\vec{v} + w_q\delta\vec{u}/\psi + \vec{\nu} \times \delta\vec{u}/\psi)^2 + \frac{1}{\psi}\, (1 - \nu^2)\, \delta\vec{u}^2$$

$$+ \frac{1}{\psi}\, (\vec{\nu}.\delta\vec{u})^2 + \frac{1}{2p}\, [w_p - (1 + w_q^2)/\psi]\, (\vec{u} \times \delta\vec{u})^2$$

$$+ \frac{1}{2r}\, (w_r - \psi)\, (\vec{v} \times \delta\vec{v})^2 + Q'',$$

$$Q'' = \delta w_p \delta p + \delta w_q \delta q + \delta w_r \delta r + \left[w_p - (1 + w_q^2)/\psi \right] \delta p^2/2p$$
$$+ (w_r - \psi) \, \delta r^2/2r$$

the convexity of Q is ensured with

$$0 < (1 + w_q^2)/w_p \leq \psi \leq w_r, \quad Q'' \geq 0.$$

We give two applications.

D) Nonlinear electrodynamics

From the electromagnetic field tensor two independent invariants can be constructed by contracting the tensor with itself and with its dual

$$Q = \gamma - \alpha, \quad R = \beta.$$

By (90) - (92) and since $L = L(Q, R)$,

$$\phi = L_Q > 0$$
$$-L_Q Q' = L_Q(\delta Q \delta L_Q + \delta R \delta L_R) + B^2 \delta L_Q^2 + 2R \delta L_Q \delta L_R + E^2 \delta L_R^2 \leq 0$$

which is just the condition that the electric field must not exceed the smallest of the absolute fields, ζ_1, ζ_2 [187].

For the Born-Infeld Lagrangian

$$L = \sqrt{-R^2 + k(2Q + k)}, \quad k = \text{const.} > 0$$

these fields are both equal to k and therefore

$$E^2 \leq k.$$

E) Nonlinear electromagnetic materials

The evolution equations

$$\partial_t \vec{D} - \text{curl } \vec{H} = 0, \quad \partial_t \vec{B} + \text{curl } \vec{E} = 0$$

are completed by the constitutive relations

$$\vec{D} = \varepsilon(E^2)\vec{E}, \quad \vec{B} = \mu(H^2)\vec{H}.$$

The Lagrangian is

$$L(\alpha, \gamma) = \frac{1}{2} \int d(B^2)/\mu(H^2) - \frac{1}{2} \int \varepsilon(E^2) d(E^2)$$

and (91), (92) read

(93)
$$0 < \frac{1}{\mu} \leq \phi \leq \varepsilon,$$

$$Q' = (2\varepsilon' E^2 + \varepsilon - \phi) \, \delta\alpha^2/E^2 + (2\mu' H^2 + \mu - \phi^{-1}) \, \delta\gamma^2/\mu^2 H^2 (2\mu' H^2 + \mu)^2 \geq 0.$$

Thus

(94)
$$0 < \frac{1}{2\mu' H^2 + \mu} \leq \phi \leq 2\varepsilon' E^2 + \varepsilon.$$

From (93) and (94) follow the conditions [188]

$$0 < \frac{1}{\mu - 2H^2\mu'_-} \leq \varepsilon - 2E^2\varepsilon'_-, \quad \varepsilon'_- := \frac{1}{2}(|\varepsilon'| - \varepsilon').$$

Acknowledgements

These lectures reflect a long collaboration with Italy which began at a CIME course in Bressanone/Brixen. We are grateful to the CIME and to the director of the course Professor T. Ruggeri. We are indebted to Dr. Ines Quandt, Berlin ; her lecture notes were a useful reminder in the preparation of the text. We thank Mrs. Irene Fontaine-Gilmour for typing and finding the way through the annotated manuscript.

References

0. **E. Infeld and G. Rowlands**, *Nonlinear Waves, Solitons and Chaos*, Cambridge University Press, Cambridge, New York (1st ed. 1990; 2nd revised ed. 1992).

1. **R. Courant and D. Hilbert**, *Methods of Mathematical Physics*, Vol.II, Interscience Publ., New York, London, Sydney (1962).

2. **A. Jeffrey and T. Taniuti**, *Non-linear Wave Propagation, with Applications to Physics and Magnetohydrodynamics*, Academic Press, New York, London (1964).

3. **A. Jeffrey**, *Quasilinear hyperbolic systems and waves*, Pitman Publishing, London, San Francisco, Melbourne (1976).

4. **A. Jeffrey**, *Lectures on nonlinear wave propagation*, in *Wave Propagation*, Corso CIME, Bressanone, 1980, Ed. by G. Ferrarese, Liguori, Napoli (1982).

5. **C. Dafermos**, *Hyperbolic systems of conservation laws*, in *Systems of nonlinear partial differential equations*. Ed. by J. Ball. NATO ASI series C, N°111, Reidel Publ. Co. (1983) pp.25-70.

6. **J. Smoller**, *Shock waves and reaction-diffusion equations*, Springer-Verlag, New York, Heidelberg, Berlin (1983).

7. **D. Serre**, *Systèmes de lois de conservation*, I & II, Diderot, Paris (1996).

8. **B. Sévennec**, *Géométrie des systèmes hyperboliques de lois de conservation*, Mémoire 56, Soc. Math. Fr. (1994).

9. **P. D. Lax**, *The multiplicities of eigenvalues*, Bull. Amer. Math. Soc.. **6** (1982) p.213.

10. **S. Friedland, J. W. Robbin and J. H. Sylvester**, *On the crossing rule*, Comm. Pure Appl. Math., **37** (1984) p.19.

11. **J. Hadamard**, *Leçons sur la propagation des ondes*, Hermann, Paris (1903).

12. **C. Reid**, *Courant in Göttingen and New York. The Story of an Improbable Mathematician*, Springer-Verlag, New York (1976) p.279.

13. **R. Courant and P. D. Lax**, *The propagation of discontinuities in wave motion*, Proc. Natl. Acad. Sci. U.S., **42** (1956) pp.872-76.

14. **C. Cattaneo**, *Elementi di teoria della propagazione ondosa*. Lezioni raccolte da S. Pluchino. Quaderni dell'Unione Matematica Italiana 20, Pitagora Editrice, Bologna (1981).

15. **A. Jeffrey**, *The development of jump discontinuities in nonlinear hyperbolic systems of equations in two independent variables*, Arch. Ratl. Mech. Analys., **14** (1963) pp.27-37.

16. **G. Boillat and T. Ruggeri**, *On the evolution law of weak discontinuities for hyperbolic quasi-linear systems*, Wave Motion, **1** (1979) pp.149-152.

17. **P. D. Lax**, *Asymptotic solutions of oscillatory initial value problem*, Duke Math. Journ., **24** (1957) pp.627-646.

18. **D. Ludwig**, *Exact and asymptotic solutions of the Cauchy problem*, Comm. Pure Appl. Math., **13** (1960) pp.473-508.

19. **Y. Choquet-Bruhat**, *Ondes asymptotiques pour un système d'équations aux dérivées partielles non linéaires*, J. Maths. pure appl., **48** (1969) pp.117-158 ; *Ondes Asymptotiques*, in *Wave Propagation*, Corso CIME. Bressanone, 1980 (op. cit.) pp.99-165.

20. **A. M. Anile and A. Greco**, *Asymptotic waves and critical time in general relativistic magnetohydrodynamics*, Ann. Inst. Henri Poincaré, **XXIX** (1978) p.257.

21. **J. K. Hunter and J. B. Keller**, *Weakly nonlinear high frequency waves*, Comm. Pure Appl. Math., **XXXVI** (1983) pp.547-569.

22. **G. Boillat**, *Ondes asymptotiques non linéaires*, Ann. Mat. pura ed appl., **CXI** (1976) pp.31-44.

23. **D. Serre**, *Oscillations non-linéaires hyperboliques de grande amplitude: DIM* ≥ 2, in *Nonlinear variational problems and partial differential equations*. Proc. of the 3rd conf., Isola d'Elba, 1990. Ed. by A. Marino et al., Pitman Res. Notes Math., Ser.320 (1995) pp.245-294.

24. **R. Courant and K. O. Friedrichs**, *Supersonic flow and shock waves*, Interscience Publ., New York (1948).

25. **P.D. Lax**, *The initial value problem for nonlinear hyperbolic equations in two independent variables*, Ann. Math. Studies, Princeton, **33** (1954) pp.211-29.

26. **P.D. Lax**, *Hyperbolic systems of conservation laws*, II, Comm. Pure Appl. Math., **10** (1957) pp.537-566.

27. **G. Boillat**, *Sur la croissance des ondes simples et l'instabilité de chocs caractéristiques des systèmes hyperboliques avec application à la discontinuité de contact d'un fluide*, C. R. Acad. Sci. Paris, **284** A (1977) pp.1481-1484.

28. **G. Boillat**, *On nonlinear plane waves*, in 8th Internat. Conf. on *Waves and Stability in Continuous Media*, Palermo, 1995. Ed. by A.M. Greco and S. Rionero (to appear in a special issue of Rend. Circ. Mat. Palermo).

29. **G. Velo and D. Zwanziger**, *Propagation and quantization of Rarita-Schwinger waves in an external electromagnetic potential*, Phys. Rev., **186** (1969) pp.1337-1341.

30. **G. Boillat**, *Exact plane wave solution of Born-Infeld electrodynamics*, Lett. N. Cim., serie 2, **4** (1972) pp.274-276.

31. **G. Boillat**, *Covariant disturbances and exceptional waves*, J. Math. Phys., **14** (1973) pp.973-976.

32. **H. C. Tze**, *Born duality and strings in hadrodynamics and electrodynamics*, N. Cim., **22** A (1974) pp.507-526.

33. **H. Freistühler**, *Linear degeneracy and shock waves*, Math. Z., **207** (1991) pp.583-596.

34. **G. Boillat**, *Chocs caractéristiques*, C.R. Acad. Sci. Paris, **274** A (1972) pp.1018-1021.

35. **G. Boillat and A. Muracchini**, *Chocs caractéristiques de croisement*, C.R. Acad. Sci. Paris, **310** I (1990) pp.229-232.

36. **G. Boillat**, *Le cône critique et le champ scalaire*, C.R. Acad. Sci. Paris, **260** (1965) pp.2427-2429.

37. **W. Heisenberg and H. Euler**, Zeitschrift für Physik, **98** (1936) p.714.

38. **G. B. Witham**, *Linear and Nonlinear Waves*, John Wiley & Sons, New York, London, Sydney, Toronto (1974).

39. **Th. von Kármán**, *Compressibility effects in aerodynamics*, J. Aeron. Sci., **8** (1941) pp.337-356.

40. **E. Carafoli**, *High Speed Aerodynamics*, Pergamon Press (1956).

41. **Y. Choquet-Bruhat**, *Théorèmes d'existence globaux pour des fluides ultra-relativistes*, C.R. Acad. Sci. Paris, **319** I (1994) pp.1337-1342.

42. **T. Ruggeri, A. Muracchini and L. Seccia**, *Shock waves and second sound in a rigid heat conductor : a critical temperature for NaF and Bi*, Phys. Rev. Lett., **64** (1990) p.2640 ; *Continuum approach to phonon gas and shape of second sound via shock waves theory*, N. Cim., **16** D (1994) pp.15-44.

43. **A. Donato and T. Ruggeri**, *Onde di discontinuità e condizioni di eccezionalità per materiali ferromagnetici*, Rend. Accad. Naz. Lincei, Serie VIII, **LIII** (1972) pp.289-294.

44. **G. Boillat**, *La propagation des ondes*, Gauthier-Villars, Paris.

45. **T. Ruggeri**, *Sulla propagazione di onde elettromagnetiche di discontinuità in mezzi non lineari*, Ist. Lombardo (Rend. Sc.) A **107** (1973) pp.283-297.

46. **T. Taniuti**, *On wave propagation in non-linear fields*, Suppl. Progr. Theor. Phys., N°9 (1959) pp.69-128.

47. **M. Born**, Proc. Roy. Soc. London, Ser. A, **143** (1933) p.410.

48. **W. Heisenberg**, Zs. f. Phys., **133** (1952) p.79 ; **126** (1949) p.519 ; **113** (1939) p.61.

49. **B. M. Barbishov and N. A. Chernikov**, *Solution of the two plane wave scattering problem in a nonlinear scalar field theory of the Born-Infeld type*, Sov. Phys. J. E. T. P., **24** (1966) pp.437-442.

50. **I. Imai**, Progr. Theor. Phys., **2** (1947) p.97.

51. **T. Taniuti**, *On the Heisenberg's non-linear meson equation*, Progr. Theor. Phys., **14** (1955) pp.408-409.

52. **A. Greco**, *On the exceptional waves in relativistic magnetohydrodynamics*, Rend. Accad. Naz. Lincei, serie VIII, **LII** (1972) pp.507-512.

53. **M. Born and L. Infeld**, *Foundations of the new field theory*, Proc. Roy. Soc. London, A **144** (1934) pp.425-451 ; **M. Born**, *Structure atomique de la matière*, Coll. U. Armand Colin, Paris (1971) ; *Atomic Physics*, Blackie and Son, Ltd., London.

54. **J. Plebański**, *Lectures on non-linear electrodynamics*, given at the Niels Bohr Institute and NORDITA in 1968, NORDITA, Copenhagen (1970). **I. Bialynicki-Birula**, *Nonlinear Electrodynamics: variations on a theme by Born and Infeld* in *Quantum Theory and Particles and Fields*. B. Jancewicz and J. Lukierski eds. World Scientific, Singapore (1983).

55. **J. Naas and H.L. Schmid**, *Mathematisches Wörterbuch mit Einbeziehung der theoretischen Physik*, Akademie Verlag, Berlin und B.G. Teubner, Stuttgart (1961).

56. **A. Einstein, H. and M. Born**, *Briefwechsel, 1916-1955*, kommentiert von Max Born. Geleitwort von Bertrand Russell, Vorwort von Werner Heisenberg. Nymphenburger Verlagshandlung, München, 1969 ; H. and M. Born, *Der Luxus des Gewissens*, Ibid., 1969.

57. **G. Boillat**, *Nonlinear electrodynamics. Lagrangians and equations of motion*, J. Math. Phys., **11** (1970) pp.941-951 ; *Born-Infeld particle ; small scale phenomena*, Lett. N. Cim., **4** (1970) pp.773-778 ; *Exact plane-wave solution of Born-Infeld electrodynamics*, Ibid., **4** (1972) pp.274-276 ; *Shock relations in nonlinear electrodynamics*, Phys. Lett., **40 A** (1972) pp.9-10.

58. **A. Strumia** *Einstein equations, relativistic strings and Born-Infeld electrodynamics*, N. Cim., **110 B** (1995) pp.1497-1504.

59. **J. H. Olsen and A. S. Shapiro**, *Large-amplitude unsteady flow in liquid-filled elastic tubes*, J. Fluid Mech., **29** (1967) pp.513-538.

60. **W. A. Green** *The growth of plane discontinuities propagating into a homogeneously deformed elastic material*, Arch. Ratl. Mech. Analys., **16** (1964) pp.79-88.

61. **N. H. Scott**, *Acceleration waves in incompressible elastic solids*, Quart. J. Mec. Appl. Math., **29** (1976) pp.259-310.

62. **A. Jeffrey and M. Teymur**, *Formation of shock waves in hyperelastic solids*, Acta Mech., **20** (1974) pp.133-149.

63. **L. Lustman**, *A note on nonbreaking waves in hyperelastic materials*, Stud. Appl. Math., **63** (1980) pp.147-154.

64. **A. Donato**. *Legge di evoluzione delle discontinuità e determinazione di una classe di potenziali elastici compatibile con la propagazione di one eccezionali in un mezzo continuo sottoposto a particolari deformazioni finite*, ZAMP, **28** (1977) pp.1059-1066.

65. **G. Boillat and T. Ruggeri**, *Su alcune classi di potenziali termodinamici come conseguenza dell'esistenza di particolari onde di discontinuità nella meccanica dei continui con deformazioni finite*, Rend. Sem. Mat. Univ. Padova, **51** (1974) pp.293-304.

66. **G. Boillat and S. Pluchino**, *Onde eccezionali in mezzi iperelastici con deformazioni finite piane*, ZAMP, **35** (1984) pp.363-372.

67. **G. Grioli**, *On the thermodynamic potential for continuums with reversible transformations - some possible types*, Meccanica, **1** (1966) pp.15-20.

68. **C. Tolotti**, *Deformazioni elastiche finite : onde ordinarie di discontinuità e casi tipici di solidi elastici isotropi*, Rend. Mat. Appl., serie V, **4** (1943) pp.34-59.

69. **G. Boillat**, *Le champ scalaire de Monge-Ampère*, Det Kgl. Norske Vid. Selsk. Forh., **41** (1968) pp.78-81.

70. **G. Valiron**, *Cours d'analyse mathématique. Equations fonctionnelles. Applications*, Masson, Paris (1950).

71. **G. Boillat**, *Sur l'équation générale de Monge-Ampère à plusieurs variables*, C. R. Acad. Sci. Paris, **313** I (1991) pp.805-808.

72. **T. Ruggeri**, *Su una naturale estensione a tre variabili dell'equazione di Monge-Ampère*, Accad. Naz. dei Lincei, **LV** (1973) pp.445-449.

73. **A. Donato, U. Ramgulam and C. Rogers**, *The 3+1 dimensional Monge-Ampère equation in discontinuity wave theory : application of a reciprocal transformation*, Meccanica, **27** (1992) pp.257-262.

74. **G. Boillat**, *Sur l'équation générale de Monge-Ampère d'ordre supérieur*, C. R. Acad. Sci. Paris, **315** I (1992) pp.1211-1214.

75. **A. Donato and F. Oliveri**, *Linearization of completely exceptional second order hyperbolic conservative equations*, Applicable Analys., **57** (1995) pp.35-45.

76. **G. Ruppeiner**, *Riemannian geometry of thermodynamics and critical phenomena*, in *Advances in Thermodynamics*. Ed. by S. Sieniutycz & P. Salomon, Vol.3, Taylor & Francis, New York (1990) pp.129-159.

77. **A. Donato and G. Valenti**, *Exceptionality criterion and linearization procedure for a third order nonlinear partial differential equation*, J. Math. Analys. Appls., **186** (1994) pp.375-382.

78. **G. Boillat**, *Sur la forme générale du système de Monge-Ampère*, Conf. at the CIRAM, Università di Bologna, October 1995 (to appear).

79. **K. O. Friedrichs**, *Symmetric hyperbolic linear differential equations*, Comm. Pure Appl. Math., **7** (1954) pp.345-392.

80. **A. Fisher and D. P. Marsden**, *The Einstein evolution equations as a first order quasi-linear symmetric hyperbolic system*, Comm. Math. Phys., **28** (1972) pp.1-38.

81. **T. Kato**, *The Cauchy problem for quasi-linear symmetric hyperbolic systems*, Arch. Ratl. Mech. Analys., **58** (1975) pp.181-205 ; *Quasi-linear equations of evolution of hyperbolic type with applications to partial differential equations*, Lect. Notes in Math., N°448, Springer Verlag, New York, Berlin (1975).

82. **S. K. Godunov**, *An interesting class of quasilinear systems*, Sov. Math. Doklady, **2** (1961) pp.947-949.

83. **K. O. Friedrichs and P. D. Lax**, *Systems of conservation equations with a convex extension*, Proc. Nat. Acad. Sci. USA, **68** (1971) pp.1686-1688.

84. **B. L. van der Waerden**, *Euler als Schöpfer der Kontinuums-Mechanik*, Naturwissenschaften, **71** (1984) pp.414-417.

85. **Schweizer Lexikon 91** in sechs Bänden, Verlag Schweizer Lexikon, Luzern (1992-93) (*Euler*).

86. **S. K. Godunov**, *The problem of a generalized solution in the theory of quasilinear equations and in gas dynamics*, Russian Math. Surveys, **17** (1962) pp.145-156.

87. **P. Carbonaro**, *Exceptional relativistic gas dynamics*, Phys. Lett. A, **129** (1988) pp.372-376.

88. **A. Greco**, *Discontinuités des rayons et stricte exceptionnalité en magnéto-hydrodynamique relativiste*, Ann. Inst. Henri Poincaré, **XII** (1975) pp.217-227.

89. **A. H. Taub**, *Relativistic Rankine-Hugoniot equations*, Phys. Rev., **74** (1948) pp.328-334.

90. **A. H. Taub**, *General relativistic shock waves in fluids for which pressure equals energy density*, Comm. math. Phys., **29** (1973) pp.79-88.

91. **G. Boillat**, *A moi compte deux mots. A spelling remark*, Math. Intelligencer, **4** (1982) p.2.

92. **T. Ruggeri and A. Strumia**, *Main field and convex covariant density for quasi-linear hyperbolic systems*, Ann. Inst. Henri Poincaré, A **XXXIV** (1981) pp.65-84.

93. **G. Boillat**, *Sur l'existence et la recherche d'équations de conservation supplémentaires pour les systèmes hyperboliques*, C. R. Acad. Sci. Paris, **278** A (1974) pp.909-912.

94. **I-Shih Liu**, *Method of Lagrange multipliers for exploitation of the entropy principle*, Arch. Ratl. Mech. Analys., **46** (1972) pp.131-148.

95. **A. Strumia**, *Main field and symmetric hyperbolic form of the Dirac equation*, Lett. N. Cim., **36** (1983) pp.609-613.

96. **G. Boillat and T. Ruggeri**, *Symmetric form of nonlinear mechanics equations and entropy growth across a shock*, Acta Mech., **35** (1980) pp.271-274.

97. **V. I. Kondaurov**, *Conservation laws and symmetrization of the equations of the nonlinear theory of elasticity*, Sov. Phys. Dokl., **26** (1981) pp.234-236.

98. **C. Truesdell**, *Six lectures on modern natural philosophy*, Springer-Verlag, New York (1966) p.72.

99. **I. Müller**, *The coldness, a universal function in thermoelastic bodies*, Arch. Ratl. Mech. Analys., **41** (1971) pp.319-332.

100. **I. Müller and T. Ruggeri**, *Extended Thermodynamics*, Springer Tracts in Natural Philosophy, Vol.37, Springer-Verlag, New York, Berlin (1993).

101. **K. O. Friedrichs**, *On the laws of relativistic electro-magneto-fluid dynamics*, Comm. Pure Appl. Math., **27** (1974) pp.749-808 ; *Conservation equations and the law of motion in classical physics*, Ibid., **31** (1978) pp.123-131.

102. **G. Boillat**, *Chocs dans les champs qui dérivent d'un principe variationnel : équation de Hamilton-Jacobi pour la fonction génératrice*, C. R. Acad. Sci. Paris **283** A (1976) pp.539-542.

103. ***Encyclopaedia of Mathematics***, Kluwer Academic Publ., Dordrecht, Boston, London (1988-94).

104. **P. Roman**, *Theory of elementary particles*, North-Holland (1960).

105. **L. L. Foldy**, *Relativistic wave equations*, in D. R. Bates, *Quantum Theory*, Vol.III, *Radiation and High Energy Physics*, Academic Press, New York and London (1962) p.42.

106. **N. N. Bogoliubov and D. V. Shirkov**, *Introduction to the theory of quantized fields*, 3rd. Ed. John Wiley & Sons, New York (1980) p.41 sqq.

107. **S. De Leo**, *Duffin-Kemmer-Petiau equation on the quaternion field*, Progr. Theor. Phys., **94** (1995) pp.1109-1120.

108. **G. Boillat**, *Involutions des systèmes conservatifs*, C. R. Acad. Sci. Paris, **307** I (1988) pp.891-894.

109. **A. Lichnerowicz**, *Théories relativistes de la gravitation et de l'électromagnétisme*, Masson, Paris (1955).

110. **Y. Choquet-Bruhat**, *Problème de Cauchy pour les modèles gravitationnels avec termes de Gauss-Bonnet*, C. R. Acad. Sci. Paris, **306** I (1988) pp.445-450.

111. **C. M. Dafermos**, *Quasilinear hyperbolic systems wih involutions*, Arch Ratl. Mech. Analys., **94** (1986) pp.373-389.

112. **G. Boillat**, *Symétrisation des systèmes d'équations aux dérivées partielles avec densité d'énergie convexe et contraintes*, C. R. Acad. Sci. Paris, **295** I (1982) pp.551-554.

113. **S. K. Godunov**, *Symmetric form of the equations of magnetohydrodynamics* (in Russian), Prepr. Akad. Nauk S.S.S.R. Sib. Otd., Vychisl. Tsentr, 3 N°1 (1972) p.26.

114. **G. Boillat and S. Pluchino**, *Sopra l'iperbolicità dei sistemi con vincoli e considerazioni sul superfluido e la magnetoidrodinamica*, ZAMP **36** (1985) pp.893-900.

115. **S. Lundquist**, *Studies in magneto-hydrodynamics*, Arkiv för Fysik, **5** (1952) p.297.

116. **L. Landau and E. Lifchitz**, *Mécanique des fluides*, MIR, Moscou (1971).

117. **G. Grioli**, Ed., *Macroscopic theories of superfluids*, Internat. Meeting, Accad. Naz. dei Lincei, Rome, May 1988. Cambridge University Press (1991).

118. **G. Boillat and A. Muracchini**, *On the symmetric conservative form of Landau's superfluid equations*, ZAMP, **35** (1984) pp.282-288 ; *On a special form of the hydrodynamic superfluid equations*, Ibid., **36** (1985) pp.901-904 ; *Thermodynamic conditions for a symmetric form of superfluid equations*, N. Cim., **9** D (1987) pp.253-259.

119. **G. Boillat and T. Ruggeri**, *Hyperbolic principal subsystems : entropy, convexity and subcharacteristic conditions*, Arch. Ratl. Mech. Analys. (to appear).

120. **T.-P. Liu**, *Hyperbolic conservation laws with relaxation*, Comm. Math. Phys., **108** (1987) pp.153-175.

121. **G.-Q. Chen, C. D. Levermore and T.-P. Liu**, *Hyperbolic conservative laws with stiff relaxation terms and entropy*, Comm. Pure Appl. Math., **67** (1994) pp.787-830.

122. **W. Weiss**, *Hierarchie der Erweiterten Thermodynamik*, Dissertation TU Berlin (1990).

123. **C. Cercignani and A. Majorana**, *Analysis of thermal and shear waves according to the B. G. K. kinetic model*, ZAMP, **36** (1985) pp.699-711.

124. **H. Ott**, *Lorentz-Transformation der Wärme und der Temperatur*, Zs. f. Phys., **175** (1963) pp.70-104.

125. **H. Arzeliès**, *Relativistic transformation of temperature and some other thermodynamical quantities* (in French), N. Cim., **35** (1965) pp.792-804.

126. **C. Møller**, *The Theory of Relativity*, 2nd Ed., Clarendon Press, Oxford (1972).

127. **J. Putterman**, *Superfluid hydrodynamics*, North Holland, Amsterdam (1974).

128. **J. R. Dorroh**, *Nonlinear symmetric first-order systems*, J. Math. Analys. Appls., **91** (1983) pp.523-526.

129. **G. Boillat**, *On symmetrization of partial differential systems*, Applicable Analys., **57** (1995) pp.17-21.

130. **G. Boillat**, *Limitation des vitesses de choc quand la densité d'énergie est convexe et les contraintes involutives*, C. R. Acad. Sci. Paris, **297** I (1983) pp.141-143.

131. **G. Boillat**, *Evolution des chocs caractéristiques dans les champs dérivant d'un principe variationnel*, J. Maths. Pures et Appl., **56** (1977) pp.137-147.

132. E. Hölder, *Historischer Überblick zur mathematischen Theorie von Unstetig-keitswellen seit Riemann und Christoffel*, in *E. B. Christoffel : The Influence of his Work on Mathematics and the Physical Sciences*, Internat. Symp., Aachen, 1979 (1981).

133. P. Germain, *Shock waves, jump relations and structure*, Adv. in Appl. Mech. Ed. by Chia-Shun Yih, Vol.12, Academic Press, New York (1972) pp.131-144.

134. P. D. Lax, *Hyperbolic systems of conservation laws and the mathematical theory of shock waves*, Regional Conf. Series in Appl. Math., N°11, SIAM, Philadelphia (1973).

135. C. M. Dafermos and R. J. Diperna, *The Riemann problem for certain classes of hyperbolic systems of conservation laws*, J. Diff. Eqs., **20** (1976) pp.90-114.

136. M. Berger and M. Berger, *Perspectives in Nonlinearity*, W.A. Benjamin Inc., New York (1968) p.137.

137. G. Boillat and T. Ruggeri, *Limite de la vitesse des chocs dans les champs à densité d'énergie convexe*, C. R. Acad. Sci. Paris, **289** A (1979) pp.257-258.

138. G. Boillat and A. Strumia, *Limitation des vitesses d'onde et de choc quand la densité relativiste d'entropie (ou d'énergie) est convexe*, Ibid., **307** I (1988) pp.111-114.

139. B. L. Keyfltz and H. C. Kranzer, *A system of non-strictly hyperbolic conservation laws arising in elasticity theory*, Arch. Ratl. Mech. Analys., **72** (1980) pp.219-241.

140. C. M. Dafermos, *Admissible wave fans in nonlinear hyperbolic systems*, Ibid., **106** (1989) pp.243-260.

141. T.-P. Liu, *Admissible solutions of hyperbolic conservation laws*, Memoirs Am. Math. Soc., **240** (1981) pp.1-78.

142. G. Boillat, *Hamilton-Jacobi equation for shocks with a convex entropy density*, in 7th Conf. on *Waves and Stability in Continuous Media*, Bologna, 1993. Ed. by S. Rionero & T. Ruggeri. Quaderno CNR. Series in Adv. in Math. & Appl. Sciences, Vol.23, World Scientific, Singapore, New Jersey, London, Hong-Kong (1994) pp.22-25.

143. G. Boillat, *Chocs avec contraintes et densité d'énergie convexe*, C. R. Acad. Sci. Paris, **295** I (1982) pp.747-750.

144. **G. Boillat**, *De la nature des chocs*, in Onde e stabilità nei mezzi continui, Cosenza, giugno 1983. Quaderni del CNR, Catania (1986) pp.35-39.

145. **G. Boillat**, *De la vitesse de choc considérée comme valeur propre*, C. R. Acad. Sci. Paris, **302** I (1986) pp.555-557.

146. **P. D. Lax**, *Shock waves and entropy*, in Contribution to non linear functional analysis. Ed. by Zarantonello, Academic Press, New York (1971).

147. **C. Dafermos**, *Generalized characteristics in hyperbolic systems of conservation laws*, Arch. Ratl. Mech. Analys., **107** (1989) p.127.

148. **D. Fusco**, *Alcune considerazioni sulle onde d'urto in fluidodinamica*, Atti Sem. Mat. Fis. Univ. Modena, **28** (1979) p.223.

149. **N. Virgopia and F. Ferraioli**, *On the shock wave generating function in a single mixture of gases*, N. Cim., D **7** (1988) p.151.

150. **A. Strumia**, *A detailed study of entropy jump across shock waves in relativistic fluid dynamics*, N. Cim., B **92** (1986) p.91.

151. **W. Dreyer and S. Seelecke**, *Entropy and causality as criteria for the existence of shock wave in low temperature ideal conduction*, Cont. Mech. Thermodyn., **4** (1992) pp.23-36.

152. **L. Seccia**, *Shock wave propagation and admissibility criteria in a non linear dielectric medium*, Ibid., **7** (1995) pp.277-296.

153. **N. Manganaro and G. Valenti**, *On shock propagation for a compressible fluid supplemented by a generalized von Kármán law*, Atti. Sem. Mat. Fis. Univ. Modena, **XXXVIII** (1990) pp.109-122.

154. **G. Boillat and A. Muracchini**, *The structure of the characteristic shocks in constrained symmetric systems with applications to magnetohydrodynamics*, Wave Motion, **11** (1989) pp.297-307.

155. **G. Boillat**, *Chocs caractéristiques et ondes simples exceptionnelles pour les systèmes conservatifs à intégrale d'énergie ; forme explicite de la solution*, C. R. Acad. Sci. Paris, **280** A (1975) pp.1325-1328.

156. **T. Ruggeri**, *On some properties of electromagnetic shock waves in isotropic nonlinear materials*, Boll. Un. Mat. Ital., **9** (1974) pp.513-522.

157. **G. Boillat**, *Expression explicite des chocs caractéristiques de croisement*, C. R. Acad. Sci. Paris, **312** I (1991) pp.653-656.

158. **A. Jeffrey**, *Magnetohydrodynamics*, Oliver & Boyd, London (1966).

159. **H. Cabannes**, *Theoretical magnetofluid-dynamics*, Academic Press, New York (1970).

160. **G. Boillat and A. Muracchini**, *Characteristic shocks of crossing velocities in magnetohydrodynamics*, NoDEA, **3**, (1996), pp. 217-230.

161. **G. Boillat and T. Ruggeri**, *Characteristic shocks : completely and strictly exceptional systems*,Boll. Un. Mat. Ital.. **15** A (1978) pp.197-204.

162. **P. D. Lax**, *Development of singularities of solutions of nonlinear hyperbolic partial differential equations*, J. Math. Phys., **5** (1964) pp.611-613.

163. **G. Boillat**, *Shock relations in nonlinear electrodynamics*, Phys. Lett., **40** A (1972) pp.9-10.

164. **G. Boillat and T. Ruggeri**, *Euler equations for nonlinear relativistic strings. Introduction of a new Lagrangian*, Progr. Theor. Phys., **60** (1978) pp.1928-1929.

165. **A. Greco**, *On the strict exceptionality for a subsonic flow*, in 2° Congresso Naz. AIMETA, Università di Napoli. Facoltà di Ingegneria, **4** (1974) pp.127-134.

166. **G. Boillat**, *A relativistic fluid in which shock fronts are also wave surfaces*, Phys. Lett., **50** A (1974) pp.357-358.

167. **L. Brun**, *Ondes de choc finies dans les solides élastiques*, in *Mechanical waves in solids*. Ed. by J. Mandel & L. Brun, Springer, Vienna (1975).

168. **A. Morro**, Atti Accad. Naz. Lincei Rend., **64** (1978) p.177 ; Acta Mech., **38** (1981) p.241 ; *On stablity in the interaction between shock waves and acoustic waves*, in *Onde e stabilità nei mezzi continui*, Catania, Nov.1981. Quaderni del CNR, Catania (1982) pp.18-47.

169. **A. Strumia**, *Transmission and reflexion of a discontinuity wave through a characteristic shock in nonlinear optics*, Riv. Mat. Univ. Parma, **4** (1978) p.15.

170. **G. Boillat and T. Ruggeri**, *Reflection and transmission of discontinuity waves through a shock wave. General theory including also the case of characteristic shocks*, Proc. Roy. Soc. Edinburgh, **83** A (1979) pp.17-24.

171. **A. Donato and D. Fusco**, *Nonlinear wave propagation in a layered half-space*, Int. J. Nonlinear Mechanics, **15** (1980) pp.497-503.

46

172. **A. M. Anile and S. Pennisi**, *On the mathematical structure of the relativistic magnetofluid-dynamics*, Ann. Inst. Henri Poincaré, **46** (1987) p.27.

173. **A. M. Anile**, *Relativistic fluids and magneto-fluids*, Cambridge University Press, 1989.

174. **G. Boillat and T. Ruggeri**, *Wave and shock velocities in relativistic magnetohydrodynamics compared with the speed of light*, Continuum Mech. Thermodyn., **1** (1989) pp.47-52.

175. **A. Lichnerowicz**, *Ondes de choc, ondes infinitésimales et rayons en hydrodynamique et magnétohydrodynamique relativistes*, in *Relativistic Fluid Dynamics*, Corso CIME, Bressanone, 1970. Ed. by C. Cattaneo, Cremonese, Roma (1971) ; *Shock waves in relativistic magnetohydrodynamics under general assumptions*, J. Math. Phys., **17** (1976) p.2125.

176. **W. Israel**, *Relativistic theory of shock waves*, Proc. Roy. Soc., **259** A (1960) p.129.

177. **A. Lichnerowicz**, *Magnetohydrodynamics : waves and shock waves in curved space-time*. Mathematical Studies, 14. Kluwer Academic Publ. Group, Dordrecht (1994).

178. **Y. Choquet-Bruhat**, *Fluides chargés non abéliens de conductivité infinie*, C. R. Acad. Sci. Paris, **314** I (1992) pp.87-91 ; *Hydrodynamics and magneto-hydrodynamics of Yang-Mills fluids*, in 7th Conf. on *Waves and Stability in continuous media*, Bologna (1993) (op. cit.).

179. **S. Pennisi**, *A covariant and extended model for relativistic magnetofluid-dynamics*, Ann. Inst. Henri Poincaré, **58** (1993) pp.343-361.

180. **G. Boillat**, *Sur l'élimination des contraintes involutives*, C. R. Acad. Sci. Paris, **318** I (1994) pp.1053-1058.

181. **C. Cattaneo**, *Sulla conduzione del calore*, Atti. Sem. Mat. Fis. Univ. Modena, **3** (1948) p.1.

182. **T. Ruggeri**, *Struttura dei sistemi alle derivate parziali compatibili con un principio di entropia e termodinamica estesa*, Suppl. Boll. Un. Mat. Ital., **4** (1985) p.261 ; *Thermodynamics and symmetric hyperbolic systems*, Rend. Sem. Mat. Univ. Torino, special issue : *Hyperbolic equations*, **167** (1987).

183. **A. Morro and T. Ruggeri**, *Non-equilibrium properties of solids obtained from second-sound measurements*, J. Phys. C : Solid State Phys., **21** (1988) p.1743.

184. **F. Franchi and A. Morro**, *Global existence and asymptotic stability in nonlinear heat conduction*, J. Math. Analys. Appls., **188** (1994) pp.590-609.

185. **G. B. Nagy, O. E. Ortiz and O. A. Reula**, *The behavior of hyperbolic heat equations' solutions near their parabolic limits*, J. Math. Phys., **35** (1994) pp.4334-4356.

186. **A. M. Anile, S. Pennisi and M. Sammartino**, *A thermodynamical approach to Eddington factors*, J. Math. Physics, **32** (1991) pp.544-550.

187. **G. Boillat**, *Convexité et hyperbolicité en électrodynamique non linéaire*, C. R. Acad. Sci. Paris, **290** A (1980) pp.259-261.

188. **G. Boillat and A. Giannone**, *A constraint on wave, radial and shock velocities in non-linear electromagnetic materials*, ZAMP, **40** (1989) pp.285-289.

Entropy and the Stability of Classical Solutions of Hyperbolic Systems of Conservation Laws

C.M. Dafermos[1]
Lefschetz Center for Dynamical Systems
Brown University
Providence, RI 02912 USA

1. Introduction

It is a tenet of Continuum Physics that the Second Law of thermodynamics is essentially a statement of stability. The Second Law manifests itself in the presense of companion balance laws, to be satisfied identically, as equalities, by classical solutions, and to be imposed as inequality thermodynamic admissibility constraints on weak solutions of the systems of balance laws. In these lecture notes we investigate the implications of entropy inequalities on the stability of classical solutions.

It will be shown that when the system of balance laws is endowed with a companion balance law induced by a convex entropy, the initial-value problem is locally well-posed in the context of classical solutions: Sufficiently smooth initial data generate a classical solution defined on a maximal time interval, of finite or infinite duration. Moreover, this solution is unique and depends continuously on the initial data, not only within the class of classical solutions but even within the broader class of weak solutions .that satisfy the companion balance law as an inequality admissibility constraint. It will further be demonstrated that the same conclusion holds even when the entropy is convex only in the direction of a certain cone in state space, provided that the system of balance laws is equipped with special companion balance laws, called involutions, whose presence compensates for the lack of convexity in complementary directions.

2. The Initial-Value Problem

We focus the investigation on homogeneous hyperbolic systems of conservation laws,

$$(2.1) \qquad \partial_t U(x,t) + \sum_{\alpha=1}^{m} \partial_\alpha G_\alpha(U(x,t)) = 0 \ .$$

In (2.1), x takes values in $I\!\!R^m$ and t in $[0,\infty)$; U takes values in some open subset $\mathcal{O}$ of $I\!\!R^n$ and G_α, $\alpha = 1, \cdots, m$, are given smooth functions from $\mathcal{O}$ to $I\!\!R^n$. Hyperbolicity means that for every fixed $U \in \mathcal{O}$ and $\nu \in S^{m-1}$ the $n \times n$ matrix

$$(2.2) \qquad \Lambda(\nu; U) = \sum_{\alpha=1}^{m} \nu_\alpha \mathcal{D}G_\alpha(U)$$

[1]Partially supported by the National Science Foundation under grant # DMS-9500574 by the Army Reasearch Office under contract # DAAH04-93-G-0198, and by the Office of Naval Reasearch under contract # N00014-92-J-1481.

has real eigenvalues $\lambda_1(\nu; U), \cdots, \lambda_n(\nu; U)$ and n linearly independent eigenvectors $R_1(\nu; U), \cdots, R_n(\nu; U)$.

With (2.1) we associate initial conditions

$$(2.3) \qquad\qquad U(x, 0) = U_0(x) \ , \ x \in I\!\!R^m \ ,$$

where U_0 is a given bounded measurable function from $I\!\!R^m$ to $\mathcal{O}$. A *classical solution* of the initial-value problem (2.1), (2.3) on a time interval $[0, T)$, is a function U from $I\!\!R^m \times [0, T)$ to $\mathcal{O}$ which is Lipschitz continuous on any compact subset of $I\!\!R^m \times [0, T)$ and satisfies (2.1) almost everywhere on $I\!\!R^m \times [0, T)$ and (2.3) for every x in $I\!\!R^m$. A *weak solution* of (2.1), (2.3) on $[0, T)$ is a locally bounded measurable function U from $I\!\!R^m \times [0, T)$ to $\mathcal{O}$ which satisfies

$$(2.4) \qquad \int_0^T \int_{I\!\!R^m} [\partial_t \phi U + \sum_{\alpha=1}^m \partial_\alpha \phi G_\alpha(U)] dx dt + \int_{I\!\!R^m} \phi(x, 0) U_0(x) dx = 0$$

for every Lipschitz continuous test function ϕ on $I\!\!R^m \times [0, T)$, with compact support. In particular, every weak solution satisfies (2.1) in the sense of distributions and any weak solution which is locally Lipschitz continuous is necessarily a classical solution.

It turns out that weak solutions are endowed with some continuity with respect to t:

Theorem 2.1 *Any bounded weak solution U of (2.1), (2.3) on $[0, T)$ may be normalized so that the map $t \mapsto U(\cdot, t)$ from $[0, T)$ to L^∞ weak* is continuous. As $t \downarrow 0, U(\cdot, t)$ converges to $U_0(\cdot)$ in L^∞ weak*.*

Proof. Fix $0 < s < \tau < T$. Let $\chi \in C_0^\infty(I\!\!R^m)$. For positive small δ and ε, write (2.4) with $\phi(x, t) = \chi(x)\theta(t)$, where $\theta(t) = 0$ for $0 \le t < s - \delta$, $\theta(t) = \delta^{-1}(t - s) + 1$ for $s - \delta \le t < s$, $\theta(t) = 1$ for $s \le t < \tau, \theta(t) = \varepsilon^{-1}(\tau - t) + 1$ for $\tau \le t < \tau + \varepsilon$, and $\theta(t) = 0$ for $\tau + \varepsilon \le t < T$. This yields

$$(2.5) \qquad \frac{1}{\delta} \int_{s-\delta}^s \int_{I\!\!R^m} \chi(x) U(x, t) dx dt - \frac{1}{\varepsilon} \int_\tau^{\tau+\varepsilon} \int_{I\!\!R^m} \chi(x) U(x, t) dx dt$$

$$+ \int_s^\tau \int_{I\!\!R^m} \sum_{\alpha=1}^m \partial_\alpha \chi(x) G_\alpha(U(x, t)) dx dt = O(\delta) + O(\varepsilon).$$

It follows that we may modify U on a set of measure zero so that

$$(2.6) \qquad \int_{I\!\!R^m} \chi(x) U(x, \tau) dx = \lim_{\varepsilon \to 0} \frac{1}{2\varepsilon} \int_{\tau-\varepsilon}^{\tau+\varepsilon} \int_{I\!\!R^m} \chi(x) U(x, t) dx dt$$

holds for any $\tau \in (0, T)$ and all χ in $L^1(I\!\!R^m)$. This normalization renders the map $t \mapsto U(\cdot, t)$ continuous in L^∞ weak*.

Next write (2.4) with $\phi(x, t) = \chi(x)\theta(t)$, where, as before, $\chi \in C_0^\infty(I\!\!R^m)$ but now $\theta(t) = 1 - \varepsilon^{-1}t$, for $0 \le t < \varepsilon$, and $\theta(t) = 0$, for $\varepsilon \le t < T$, to get

$$(2.7) \qquad -\frac{1}{\varepsilon} \int_0^\varepsilon \int_{I\!\!R^m} \chi(x) U(x, t) dx dt + \int_{I\!\!R^m} \chi(x) U_0(x) dx = O(\varepsilon) \ ,$$

whence it follows that, as $t \downarrow 0, U(\cdot, t)$ converges to $U_0(\cdot)$ in L^∞ weak*. This completes the proof.

In view of the above result one may associate with each weak solution U a *trajectory* $U(\cdot, t)$ in L^∞ weak*.

3. Proper Entropies and Admissible Solutions

We assume that our system of conservation laws (2.1) is endowed with a designated entropy η with associated entropy flux $(q_1, \cdots, q_m)$ so that

$$(3.1) \qquad Dq_\alpha(U) = D\eta(U)DG_\alpha(U) \ , \quad \alpha = 1, \cdots, m \ ,$$

$$(3.2) \qquad D^2\eta(U)DG_\alpha(U) = DG_\alpha(U)^T D^2\eta(U) \ , \quad \alpha = 1, \cdots, m.$$

We shall employ this entropy-entropy flux pair to weed out undesirable weak solutions. We shall call a weak solution U of (2.1), (2.3), on $[0, T)$, *admissible* if it satisfies the inequality

$$(3.3) \qquad \partial_t \eta(U(x, t)) + \sum_{\alpha=1}^m \partial_\alpha q_\alpha(U(x, t)) \leq 0 \ ,$$

in the sense that

$$(3.4) \qquad \int_0^T \int_{I\!\!R^m} [\partial_t \psi \eta(U) + \sum_{\alpha=1}^m \partial_\alpha \psi q_\alpha(U)] dx dt + \int_{I\!\!R^m} \psi(x, 0)\eta(U_0(x)) dx \geq 0 \ ,$$

for all nonnegative, Lipschitz continuous test functions ψ on $I\!\!R^m \times [0, T)$, with compact support. In particular, (3.3) holds in the sense of distributions. By virtue of (3.1), every classical solution of (2.1), (2.3) satisfies (3.4) and (3.3) identically, as equalities. Therefore, any classical solution is admissible.

Systems of balance laws from Continuum Physics, are equipped with such an entropy-entropy flux pair, with the induced inequality (3.3) expressing, explicitly or implicitly, the Second Law of thermodynamics. This provides the motivation for the notion of admissibility of weak solutions, introduced above, and also suggests the natural conditions on the special entropy η:

Definition 3.1 An entropy η for the system of conservation laws (2.1) is *proper* if along the trajectory $U(\cdot, t)$ of any bounded weak solution of (2.1), (2.3) the map $t \mapsto \eta(U(\cdot, t))$ from $[0, T)$ to L^∞ weak* is lower semicontinuous.

By account of Theorem 2.1, any convex entropy is necessarily proper. Convex entropies play a very important role in the theory of hyperbolic systems of conservation laws. It is well-known [G,FL,B1,RS] that the presence of a uniformly convex entropy is a necessary and sufficient condition for the system to be symmetrizable. When the entropy is uniformly convex, admissible solutions assume their initial values in a stronger sense:

Lemma 3.1 *Assume $D^2\eta(U)$ is positive definite, uniformly on compact subsets of $\mathcal{O}$. If U is any admissible weak solution of (2.1), (2.3), taking values in some convex compact subset of $\mathcal{O}$, then, for any $R > 0$,*

$$(3.5) \qquad \lim_{\varepsilon \downarrow 0} \frac{1}{\varepsilon} \int_0^\varepsilon \int_{|x| < R} |U(x, t) - U_0(x)|^2 dx dt = 0.$$

Proof. Fix $R > 0$ and $\varepsilon > 0$. Write (3.4) for the test function $\psi(x, t) = \chi(|x|)\theta(t)$ where $\chi(r) = 1$ for $0 \leq r < R$, $\chi(r) = 1 - \varepsilon^{-1}(r - R)$ for $R \leq r < R + \varepsilon$, $\chi(r) = 0$

for $R + \varepsilon \leq r < \infty$, $\theta(t) = 1 - \varepsilon^{-1}t$ for $0 \leq t < \varepsilon$, and $\theta(t) = 0$ for $\varepsilon \leq t < T$. This gives

$$(3.6) \qquad -\frac{1}{\varepsilon} \int_0^\varepsilon \int_{|x|<R} \eta(U(x,t)) dx dt + \int_{|x|<R} \eta(U_0(x)) dx + O(\varepsilon) \geq 0.$$

Combining (3.6) with Theorem 2.1 yields

$$(3.7) \quad \limsup_{\varepsilon \downarrow 0} \frac{1}{\varepsilon} \int_0^\varepsilon \int_{|x|<R} \{ \eta(U(x,t)) - \eta(U_0(x)) - D\eta(U_0(x))[U(x,t) - U_0(x)] \} dx dt \leq 0$$

whence (3.5) follows. This completes the proof.

4. Convex Entropy and the Existence of Classical Solutions

When the system of conservation laws is equipped with a uniformly convex entropy, a classical solution of the initial-value problem exists on a maximal time interval, provided the initial data are sufficiently smooth.

In what follows, the symbol r will denote a m-tuple of integers: $r = (r_1, \cdots, r_m)$. When $r \geq 0$, we put $|r| = r_1 + \cdots + r_m$ and $\partial^r = \partial_1^{r_1} \cdots \partial_m^{r_m}$. Thus ∂^r is a differential operator of order $|r|$. For the gradient operator $(\partial_1, \cdots, \partial_m)$, we shall be using the symbol ∇.

The Sobolev space H^ℓ , $\ell = 0, 1, 2, \cdots$, is defined as the completion of the space $C_0^\infty(\mathbb{R}^m)$ of smooth functions on $\mathbb{R}^m$ with compact support, under the norm

$$(4.1) \qquad \|W\|_\ell = \left[\sum_{|r| \leq \ell} \int_{\mathbb{R}^m} |\partial^r W(x)|^2 dx \right]^{\frac{1}{2}}.$$

By the Sobolev embedding theorem, for $\ell > m/2$, H^ℓ is continuously embedded in the space of continuous functions on $\mathbb{R}^m$ and

$$(4.2) \qquad \|W\|_{L^\infty} \leq a\|W\|_\ell , \text{ for any } W \in H^\ell.$$

Theorem 4.1 *Assume the system of conservation laws (2.1) is endowed with an entropy η with $D^2\eta(U)$ positive definite, uniformly on compact subsets of $\mathcal{O}$. Suppose the initial data U_0 are continuously differentiable on $\mathbb{R}^m$, take values in some compact subset of $\mathcal{O}$ and $\nabla U_0 \in H^\ell$ for some $\ell > m/2$. Then there exists $T_\infty, 0 < T_\infty \leq \infty$, and a unique continuously differentiable function U on $\mathbb{R}^m \times [0, T_\infty)$, taking values in $\mathcal{O}$, which is a classical solution of the initial-value problem (2.1), (2.3) on $[0, T_\infty)$. Furthermore,*

$$(4.3) \qquad \nabla U(\cdot, t) \in C^0([0, T_\infty); H^\ell).$$

The interval $[0, T_\infty)$ is maximal, in the sense that whenever $T_\infty < \infty$

$$(4.4) \qquad \|\nabla U(\cdot, t)\|_{L^\infty} \to \infty , \text{ as } t \uparrow T_\infty$$

and/or the range of $U(\cdot, t)$ escapes from every compact subset of $\mathcal{O}$ as $t \uparrow T_\infty$.
Proof. It is lengthy and tedious. Just an outline will be presented here, so as to illustrate the role of the convex entropy. For the details the reader may consult the monograph [M] by Majda and references therein.

Fix any open subset $\mathcal{B}$ of $I\!\!R^m$ which contains the closure of the range of U_0 and whose closure $\overline{\mathcal{B}}$ is in turn contained in $\mathcal{O}$. With positive constants ω and T, to be fixed later, we associate the class $\mathcal{F}$ of Lipschitz continuous functions V, defined on $I\!\!R^m \times [0, T]$, taking values in $\mathcal{B}$, satisfying the initial condition (2.3) and

$$(4.5) \qquad \nabla V(\cdot, t) \in L^\infty([0, T]; H^\ell) \ , \ \partial_t V(\cdot, t) \in L^\infty([0, T]; L^2 \cap L^\infty)$$

with

$$(4.6) \qquad \sup_{[0,T]} \|\nabla V(\cdot, t)\|_\ell \leq \omega \ ,$$

$$(4.7) \qquad \sup_{[0,T]} \|\partial_t V(\cdot, t)\|_{L^\infty} \leq b\omega \ , \ \sup_{[0,T]} \|\partial_t V(\cdot, t)\|_{L^2} \leq b\omega \ ,$$

where

$$(4.8) \qquad b^2 = \max_{V \in \overline{\mathcal{B}}} \sum_{\alpha=1}^{m} |DG_\alpha(V)|^2.$$

For ω sufficiently large, $\mathcal{F}$ is nonempty; for instance, $V(x, t) \equiv U_0(x)$ is a member of it.

By standard weak lower semicontinuity of norms, $\mathcal{F}$ is a complete metric space under the metric

$$(4.9) \qquad \rho(V, \overline{V}) = \sup_{[0,T]} \|V(\cdot, t) - \overline{V}(\cdot, t)\|_{L^2}.$$

Notice that, even though $V(\cdot, t)$ and $\overline{V}(\cdot, t)$ are not necessarily in L^2, $\rho(V, \overline{V}) \leq 2b\omega T < \infty$, by virtue of $V(\cdot, 0) - \overline{V}(\cdot, 0) = 0$ and (3.7).

We now linearize (2.1) about any fixed $V \in \mathcal{F}$:

$$(4.10) \qquad \partial_t U(x, t) + \sum_{\alpha=1}^{m} DG_\alpha(V(x, t))\partial_\alpha U(x, t) = 0.$$

The existence of a solution to (2.1), (2.3) on $[0, T]$ will be established by showing that

(a) When ω is sufficiently large and T is sufficiently small, the initial-value problem (4.10), (2.3) admits a solution $U \in \mathcal{F}$ on $[0, T]$.

(b) The aforementioned solution U is endowed with regularity (4.3), slightly better than (4.5) that mere membership in $\mathcal{F}$ would guarantee.

(c) For T sufficiently small, the map that carries $V \in \mathcal{F}$ to the solution $U \in \mathcal{F}$ of (4.10), (2.3) is a contraction in the metric (4.9) and thus pocesses a unique fixed point in $\mathcal{F}$, which is the desired solution of (2.1), (2.3).

In the following sketch of proof of assertion (a), above, we shall take for granted that the solution U of (4.10), (2.3), with the requisite regularity, exists and will proceed to establish that it belongs to $\mathcal{F}$. In a complete proof, one should first mollify V and the initial data, then employ the classical theory of symmetrizable, linear hyperbolic systems, and finally pass to the limit.

We fix any multi-index $r \geq 0$ of order $1 \leq |r| \leq \ell + 1$, set $\partial^r U = U_r$, and apply ∂^r to equation (4.10) to get

$$(4.11) \qquad \partial_t U_r + \sum_{\alpha=1}^m DG_\alpha(V)\partial_\alpha U_r = \sum_{\alpha=1}^m \{DG_\alpha(V)\partial^r \partial_\alpha U - \partial^r[DG_\alpha(V)\partial_\alpha U]\}.$$

The L^2 norm of the right-hand side of (4.11) may be majorized with the help of Moser-type inequalities combined with (4.2) and (4.6):

$$(4.12) \qquad \| \sum_{\alpha=1}^m \{DG_\alpha(V)\partial^r \partial_\alpha U - \partial^r[DG_\alpha(V)\partial_\alpha U]\}\|_{L^2}$$

$$\leq c\|\nabla V\|_{L^\infty}\|\nabla U\|_\ell + c\|\nabla U\|_{L^\infty}\|\nabla V\|_\ell \leq 2ac\omega\|\nabla U\|_\ell.$$

Here and below c will stand for a generic positive constant which may depend on bounds of derivatives of the G_α over $\overline{B}$ but is independent of ω and T.

Let us now multiply (4.11), from the left, by $2U_r^T D^2\eta(V)$, sum over all multi-indices r with $1 \leq |r| \leq \ell + 1$ and integrate the resulting equation over $I\!\!R^m \times [0, t]$. Note that

$$(4.13) \qquad 2U_r^T D^2\eta(V)\partial_t U_r = \partial_t[U_r^T D^2\eta(V)U_r] - 2U_r^T \partial_t D^2\eta(V)U_r.$$

Moreover, by virtue of (3.2),

$$(4.14) \qquad 2U_r^T D^2\eta(V)DG_\alpha(V)\partial_\alpha U_r = \partial_\alpha[U_r^T D^2\eta(V)DG_\alpha(V)U_r]$$

$$-2U_r^T \partial_\alpha[D^2\eta(V)DG_\alpha(V)]U_r.$$

Recall that $D^2\eta(V)$ is positive definite, uniformly on compact sets, so that

$$(4.15) \qquad U_r^T D^2\eta(V)U_r \geq \delta|U_r|^2 \ , \quad V \in \overline{B} \ ,$$

for some $\delta > 0$. Therefore, combining the above we end up with an estimate

$$(4.16) \qquad \|\nabla U(\cdot, t)\|_\ell^2 \leq c\|\nabla U_0(\cdot)\|_\ell^2 + c\omega \int_0^t \|\nabla U(\cdot, \tau)\|_\ell^2 d\tau$$

whence, by Gronwall's inequality,

$$(4.17) \qquad \sup_{[0,T]} \|\nabla U(\cdot, t)\|_\ell^2 \leq ce^{c\omega T}\|\nabla U_0(\cdot)\|_\ell^2.$$

From (4.17) follows that if ω is selected sufficiently large and T is sufficiently small, $\sup_{[0,T]} \|\nabla U(\cdot, t)\|_\ell \leq \omega$.

Then (4.10) implies $\sup_{[0,T]} \|\partial_t U(\cdot, t)\|_{L^\infty} \leq b\omega$, $\sup_{[0,T]} \|\partial_t U(\cdot, T)\|_{L^2} \leq b\omega$, with b given by (4.8). Finally, for T sufficiently small, U will take values in $\overline{B}$ on $I\!\!R^m \times [0, T]$. Thus $U \in \mathcal{F}$.

Assertion (b), namely that U is regular as in (4.3), may be established by carefully monitoring the mode of convergence of solutions of (4.10), with V mollified, to those with V in $\mathcal{F}$. This, less central, issue will not be addressed here.

Turning now to assertion (c), let us fix V and $\overline{V}$ in $\mathcal{F}$ which induce solutions U and $\overline{U}$ of (4.10), (2.3), also in $\mathcal{F}$. Thus

$$(4.18) \qquad \partial_t[U - \overline{U}] + \sum_{\alpha=1}^m DG_\alpha(V)\partial_\alpha[U - \overline{U}] = -\sum_{\alpha=1}^m [DG_\alpha(V) - DG_\alpha(\overline{V})]\partial_\alpha\overline{U}.$$

Multiply (4.18), from the left, by $2(U - \overline{U})^T D^2\eta(V)$ and integrate the resulting equation over $I\!\!R^m \times [0, t]$, $0 \leq t \leq T$. Notice that

$$(4.19) \qquad 2(U - \overline{U})^T D^2\eta(V)\partial_t(U - \overline{U}) = \partial_t[(U - \overline{U})^T D^2\eta(V)(U - \overline{U})]$$

$$-2(U - \overline{U})^T \partial_t D^2\eta(V)(U - \overline{U})$$

and also, by virtue of (3.2),

$$(4.20) \quad 2(U - \overline{U})^T D^2\eta(V) DG_\alpha(V)\partial_\alpha(U - \overline{U}) = \partial_\alpha[(U - \overline{U})^T D^2\eta(V) DG_\alpha(V)(U - \overline{U})]$$

$$-2(U - \overline{U})^T \partial_\alpha[D^2\eta(V) DG_\alpha(V)](U - \overline{U}).$$

Since $D^2\eta(V)$ is positive definite,

$$(4.21) \qquad\qquad (U - \overline{U})^T D^2\eta(V)(U - \overline{U}) \geq \delta|U - \overline{U}|^2.$$

Therefore, from the above and (4.6), (4.7), (4.2) we arrive at the estimate

$$(4.22) \qquad \|(U - \overline{U})(\cdot, t)\|_{L^2}^2 \leq c\omega \int_0^t \|(U - \overline{U})(\cdot, \tau)\|_{L^2}^2 d\tau$$

$$+c\omega \int_0^t \|(V - \overline{V})(\cdot, \tau)\|_{L^2} \|(U - \overline{U})(\cdot, \tau)\|_{L^2} d\tau.$$

Using (4.9) and Gronwall's inequality, we infer from (4.22) that

$$(4.23) \qquad\qquad \rho(U, \overline{U}) \leq c\omega T e^{c\omega T} \rho(V, \overline{V}).$$

Consequently, for T sufficiently small, the map that carries V in $\mathcal{F}$ to the solution U of (4.10), (2.3) is a contraction on $\mathcal{F}$ and thus possesses a unique fixed point U which is the unique solution of (2.1), (2.3) on $[0, T]$, in the function class $\mathcal{F}$.

Since the restriction $U(\cdot, T)$ of the constructed solution to $t = T$ belongs to the same function class as $U_0(\cdot)$, we may repeat the above construction and extend U to a larger time interval $[0, T']$. Continuing the process, we end up with a solution U defined on a maximal interval $[0, T_\infty)$ with $T_\infty \leq \infty$. Furthermore, if $T_\infty < \infty$ then the range of $U(\cdot, t)$ must escape from every compact subset of $\mathcal{O}$ as $t \uparrow T_\infty$ and/or

$$(4.24) \qquad\qquad \|U(\cdot, t)\|_\ell \to \infty \ , \quad \text{as } t \uparrow T_\infty.$$

In order to see the implications of (4.24), we retrace the steps that led to (4.16). We use again (4.12), (4.13), (4.14), and (4.15), setting $V \equiv U$, but we no longer majorize $\|\nabla U\|_{L^\infty}$ by $a\omega$. Thus, in the place of (4.16) we get

$$(4.25) \qquad \|\nabla U(\cdot, t)\|_\ell^2 \leq c\|\nabla U_0(\cdot)\|_\ell^2 + c \int_0^t \|\nabla U(\cdot, \tau)\|_{L^\infty} \|\nabla U(\cdot, \tau)\|_\ell d\tau.$$

Gronwall's inequality then implies that (4.24) cannot occur unless (4.4) does. This completes the sketch of the proof.

5. Convex Entropy and the Stability of Classical Solutions

The aim here is to show that the presence of a convex entropy guarantees that classical solutions of the initial-value problem depend continuously on the initial data, even within the broader class of admissible bounded weak solutions (see [Da,Di]).

Theorem 5.1 *Assume the system of conservation laws* (2.1) *is endowed with an entropy* η *with* $D^2\eta(U)$ *positive definite, uniformly on compact subsets of* $\mathcal{O}$. *Suppose* $\overline{U}$ *is a classical solution of* (2.1) *on* $[0, T)$, *taking values in a convex compact subset* $\mathcal{D}$ *of* $\mathcal{O}$, *with initial data* $\overline{U}_0$. *Let* U *be any admissible weak solution of* (2.1) *on* $[0, T)$, *taking values in* $\mathcal{D}$, *with initial data* U_0. *Then*

$$(5.1) \qquad \int_{|x|<R} |U(x,t) - \overline{U}(x,t)|^2 dx \leq ae^{bt} \int_{|x|<R+st} |U_0(x) - \overline{U}_0(x)|^2 dx$$

holds for any $R > 0$ *and* $t \in [0, T)$, *with positive constants* s, a, *depending solely on* $\mathcal{D}$, *and* b *that also depends on the Lipschitz constant of* $\overline{U}$. *In particular,* $\overline{U}$ *is the unique admissible weak solution of* (2.1) *with initial data* $\overline{U}_0$ *and values in* $\mathcal{D}$.

Proof. On $\mathcal{D} \times \mathcal{D}$ we define the functions

$$(5.2) \qquad h(U, \overline{U}) = \eta(U) - \eta(\overline{U}) - D\eta(\overline{U})[U - \overline{U}] \ ,$$

$$(5.3) \qquad f_\alpha(U, \overline{U}) = q_\alpha(U) - q_\alpha(\overline{U}) - D\eta(\overline{U})[G_\alpha(U) - G_\alpha(\overline{U})] \ ,$$

$$(5.4) \qquad Z_\alpha(U, \overline{U}) = G_\alpha(U) - G_\alpha(\overline{U}) - DG_\alpha(\overline{U})[U - \overline{U}] \ ,$$

all of quadratic order in $U - \overline{U}$ (recall (3.1)). Consequently, since $D^2\eta(U)$ is positive definite, uniformly on $\mathcal{D}$, there is a positive constant s such that

$$(5.5) \qquad \left[\sum_{\alpha=1}^{m} |f_\alpha(U, \overline{U})|^2\right]^{1/2} \leq sh(U, \overline{U}).$$

Let us fix any nonnegative, Lipschitz continuous test function ψ on $\mathbb{R}^m \times [0, T)$, with compact support, and evaluate h, f_α and Z_α along the two solutions $U(x,t)$, $\overline{U}(x,t)$. Recalling that U, as an admissible weak solution, must satisfy inequality (3.4), while $\overline{U}$, being a classical solution, will identically satisfy (3.4) as an equality, we deduce

$$(5.6) \qquad \int_0^T \int_{\mathbb{R}^m} [\partial_t \psi h(U, \overline{U}) + \sum_{\alpha=1}^{m} \partial_\alpha \psi f_\alpha(U, \overline{U})] dx dt + \int_{\mathbb{R}^m} \psi(x, 0) h(U_0(x), \overline{U}_0(x)) dx$$

$$\geq -\int_0^T \int_{\mathbb{R}^m} \{\partial_t \psi D\eta(\overline{U})[U - \overline{U}] + \sum_{\alpha=1}^{m} \partial_\alpha \psi D\eta(\overline{U})[G_\alpha(U) - G_\alpha(\overline{U})]\} dx dt$$

$$-\int_{\mathbb{R}^m} \psi(x, 0) D\eta(\overline{U}_0(x))[U_0(x) - \overline{U}_0(x)] dx.$$

Next we write (2.4) for both solutions U and $\overline{U}$, using components of the Lipschitz continuous vector field $\psi D\eta(\overline{U})$ as test function ϕ, to get

$$(5.7) \qquad \int_0^T \int_{\mathbb{R}^m} \{\partial_t[\psi D\eta(\overline{U})][U - \overline{U}] + \sum_{\alpha=1}^{m} \partial_\alpha[\psi D\eta(\overline{U})][G_\alpha(U) - G_\alpha(\overline{U})]\} dx dt$$

$$+ \int_{\mathbb{R}^m} \psi(x,0) D\eta(\overline{U}_0(x))[U_0(x) - \overline{U}_0(x)]dx = 0.$$

Since $\overline{U}$ is a classical solution of (2.1) and by virtue of (3.2),

(5.8)

$$\partial_t D\eta(\overline{U}) = \partial_t \overline{U}^T D^2 \eta(\overline{U}) = -\sum_{\alpha=1}^m \partial_\alpha \overline{U}^T DG_\alpha(\overline{U}) D^2 \eta(\overline{U}) = -\sum_{\alpha=1}^m \partial_\alpha \overline{U}^T D^2 \eta(\overline{U}) DG_\alpha(\overline{U})$$

so that, recalling (5.4),

(5.9)

$$\partial_t D\eta(\overline{U})[U - \overline{U}] + \sum_{\alpha=1}^m \partial_\alpha D\eta(\overline{U})[G_\alpha(U) - G_\alpha(\overline{U})]$$

$$= \sum_{\alpha=1}^m \partial_\alpha \overline{U}^T D^2 \eta(\overline{U}) Z_\alpha(U, \overline{U}).$$

Combining (5.6), (5.7) and (5.9) yields

(5.10) $$\int_0^T \int_{\mathbb{R}^m} [\partial_t \psi h(U, \overline{U}) + \sum_{\alpha=1}^m \partial_\alpha \psi f_\alpha(U, \overline{U})]dxdt + \int_{\mathbb{R}^m} \psi(x,0) h(U_0(x), \overline{U}_0(x))dx$$

$$\geq \int_0^T \int_{\mathbb{R}^m} \psi \sum_{\alpha=1}^m \partial_\alpha \overline{U}^T D^2 \eta(\overline{U}) Z_\alpha(U, \overline{U})dxdt.$$

We now fix $R > 0$, $t \in (0, T)$ and ε positive small, and write (5.10) for the test function $\psi(x, \tau) = \chi(x, \tau)\theta(\tau)$, with

(5.11)
$$\theta(\tau) = \begin{cases} 1 & 0 \leq \tau < t \\ \dfrac{1}{\varepsilon}(t - \tau) + 1 & t \leq \tau < t + \varepsilon \\ 0 & t + \varepsilon \leq \tau < T \end{cases}$$

(5.12)
$$\chi(x, \tau) = \begin{cases} 1 & 0 \leq \tau < T, \ 0 \leq |x| < R + s(t - \tau) \\ \dfrac{1}{\varepsilon}[R + s(t - \tau) - |x|] + 1 & 0 \leq \tau < T, R + s(t - \tau) \leq |x| < R + s(t - \tau) + \varepsilon \\ 0 & 0 \leq \tau < T, \ R + s(t - \tau) + \varepsilon \leq |x| < \infty \end{cases}$$

where s is the constant appearing in (5.5). The calculation gives

(5.13) $$\frac{1}{\varepsilon} \int_t^{t+\varepsilon} \int_{|x|<R} h(U(x,\tau), \overline{U}(x,\tau))dxd\tau \leq \int_{|x|<R+st} h(U_0(x), \overline{U}_0(x))dx$$

$$-\frac{1}{\varepsilon} \int_0^t \int_{R+s(t-\tau)<|x|<R+s(t-\tau)+\varepsilon} \left\{ sh(U, \overline{U}) + \sum_{\alpha=1}^m \frac{x_\alpha}{|x|} f_\alpha(U, \overline{U}) \right\} dxd\tau$$

$$-\int_0^t \int_{|x|<R+s(t-\tau)} \sum_{\alpha=1}^m \partial_\alpha \overline{U}^T D^2 \eta(\overline{U}) Z_\alpha(U, \overline{U})dxd\tau + O(\varepsilon).$$

We let $\varepsilon \downarrow 0$. The second integral on the right-hand side of (5.13) is nonnegative by account of (5.5). Using that η is convex and thereby proper (Definition 3.1), that $\overline{U}$ is Lipschitz continuous, and Theorem 2.1, we deduce

$$(5.14) \qquad \int_{|x|<R} h(U(x,t), \overline{U}(x,t))dx \leq \int_{|x|<R+st} h(U_0(x), \overline{U}_0(x))dx$$

$$-\int_0^t \int_{|x|<R+s(t-\tau)} \sum_{\alpha=1}^m \partial_\alpha \overline{U}^T D^2 \eta(\overline{U}) Z_\alpha(U, \overline{U}) dx d\tau.$$

As noted above, $h(U, \overline{U})$ and the $Z_\alpha(U, \overline{U})$ are of quadratic order in $U - \overline{U}$ and, in addition, $h(U, \overline{U})$ is positive definite, due to the convexity of η. Therefore, (5.14) in conjunction with Gronwall's inequality imply (5.1). Notice that a and s depend solely on $\mathcal{D}$ while b depends also on the Lipschitz constant of $\overline{U}$. This completes the proof.

It is remarkable that a single entropy inequality, with convex entropy, manages to weed out all but one solution of the initial-value problem, so long as a classical solution exists. However, when no classical solution exists, just one entropy inequality is no longer generally sufficient to single out any particular weak solution.

6. Nonconvex Proper Entropies and Involutions

The previous two sections have illustrated the beneficent role of convex entropies. Nevertheless, the entropy associated with systems of balance laws in Continuum Physics is not always convex. An illustrative example is provided by the system of conservation laws of isentroopic, adiabatic thermoelasticity [TN]

$$(6.1) \qquad \begin{cases} \partial_t F_{i\alpha} - \partial_\alpha v_i = 0 , & i = 1, \cdots, m, \ \alpha = 1, \cdots, m, \\[2mm] \partial_t v_i - \sum_{\alpha=1}^m \partial_\alpha T_{i\alpha}(F) = 0 , & i = 1, \cdots, m, \end{cases}$$

where v is the *velocity*, F is the *deformation gradient matrix*, and T is the *Piola-Kirchhoff stress tensor*, which is determined as the gradient of the *strain energy* function $\varepsilon(F)$:

$$(6.2) \qquad T_{i\alpha}(F) = \frac{\partial \varepsilon(F)}{\partial F_{i\alpha}} , \quad i = 1, \cdots, m, \ \alpha = 1, \cdots, m.$$

The system (6.1) is hyperbolic when the strain energy function is *rank-one convex*, i.e.,

$$(6.3) \qquad \sum_{i,j,\alpha,\beta} \frac{\partial^2 \varepsilon(F)}{\partial F_{i\alpha} \partial F_{j\beta}} \xi_i \xi_j \nu_\alpha \nu_\beta > 0 \quad , \xi, \nu \in \mathcal{S}^{m-1}.$$

The natural "entropy" for the system (6.1) is the mechanical energy

$$(6.4) \qquad \eta(F, v) = \frac{1}{2}|v|^2 + \varepsilon(F)$$

with associated entropy flux

$$(6.5) \qquad q_\alpha(F, v) = -\sum_{i=1}^m v_i T_{i\alpha}(F) \quad , \ \alpha = 1, \cdots, m.$$

Note that $\eta(F, v)$ would be convex if $\varepsilon(F)$ were convex. The hyperbolicity condition (6.3) implies that $\varepsilon(F)$ is convex, at least in directions $\xi \otimes \nu$ of rank one. This is compatible with experience with nonlinear elastic materials like rubber [TN]. However, convexity of $\varepsilon(F)$ in general directions runs contrary to physical experience and, in particular, is incompatible with the principle of *material frame indifference* [TN] which dictates that $\varepsilon(F)$ should be invariant under rigid rotations:

$$(6.6) \qquad \varepsilon(OF) = \varepsilon(F) \quad , \text{ for all orthogonal matrices } O.$$

It will be shown here that the failure of ε to be convex in certain directions is compensated by the property that solutions of the system (6.1) satisfy identically the additional conservation laws

$$(6.7) \qquad \partial_\beta F_{i\alpha} - \partial_\alpha F_{i\beta} = 0 \ , \quad i = 1, \cdots, m; \ \ \alpha, \beta = 1, \cdots, m.$$

Systems exhibiting such behavior arise quite commonly in Continuum Physics. For example, solutions of Maxwell's equations

$$(6.8) \qquad \begin{cases} \partial_t B = - \text{curl} E \\[2mm] \partial_t D = \ \text{curl} H \end{cases}$$

satisfy identically the additional conservation laws

$$(6.9) \qquad \text{div} B = 0 \ , \quad \text{div} D = 0,$$

whenever the initial data do so. Similar cases are encountered in the general theory of relativity [B2]. In view of the above, it is warranted to investigate systems of balance laws with this special structure in a general framework [B2,Da]:

Definition 6.1 The first order system

$$(6.10) \qquad \sum_{\alpha=1}^{m} M_\alpha \partial_\alpha U = 0$$

of differential equations, with M_α constant $k \times n$ matrices, $\alpha = 1, \cdots, m$, is called an *involution* of the system (2.1) of conservation laws if any (generally weak) solution of the initial-value problem (2.1), (2.3) satisfies (6.10) identically, whenever the initial data do so.

Thus (6.7) is an involution of (6.1); (6.9) is an involution of (6.8). The reader should exercise caution to distinguish involutions (6.10) which must be satisfied by all, even weak, solutions of (2.1) from conditions like (6.10) that need only hold for classical solutions. An example of the latter case is the vanishing of vorticity in smooth *irrotational flows* of Newtonian fluids: A standard calculation, that may be found in every text on hydrodynamics, shows that in any classical solution on $I\!\!R^m \times [0, T)$ of the Euler equations when the *vorticity* curl$v$ vanishes at $t = 0$ then it vanishes everywhere in space-time:

$$(6.11) \qquad \qquad \qquad \text{curl} v = 0.$$

However, (6.11) is not an involution because it does not necessarily hold for weak solutions.

A sufficient condition for (6.10) to be an involution of (2.1) is that

$$(6.12) \qquad M_\alpha G_\beta(U) + M_\beta G_\alpha(U) = 0 \ , \quad \alpha, \beta = 1, \cdots, m,$$

for any $U \in \mathcal{O}$. We shall focus our investigation here to this special case which covers, in particular, the prototypical examples (6.7) and (6.9).

With the involution (6.10) and any $\nu \in S^{m-1}$ we associate the $k \times n$ matrix

$$(6.13) \qquad N(\nu) = \sum_{\alpha=1}^{m} \nu_\alpha M_\alpha.$$

Recalling the notation (2.2), it follows from (6.12) that

$$(6.14) \qquad N(\nu)\Lambda(\nu; U) = 0$$

so, in particular, any eigenvector $R(\nu; U)$ of $\Lambda(\nu; U)$ with nonzero eigenvalue $\lambda(\nu; U)$ must lie in the kernel of $N(\nu)$. We make the simplifying assumption, valid in the prototypical examples, that, for any $\nu \in S^{m-1}$, the rank of $N(\nu)$ equals the dimension of the kernel of $\Lambda(\nu; U)$.

The premise is that, in the presence of involutions, the entropy need only be convex in the direction of a cone defined by

Definition 6.2 The *involution cone* in $\mathbb{R}^n$ of the involution (6.10) is

$$(6.15) \qquad \mathcal{C} = \bigcup_{\nu \in S^{m-1}} \ker N(\nu)$$

with $N(\nu)$ given by (6.13).

In what follows, for $p > 0$, functions on $\mathbb{R}^m$ will be called $2p$-periodic when they are periodic, with period $2p$, in each variable $x_\alpha, \alpha = 1, \cdots, m$; and $\mathcal{K}$ will denote the standard hypercube $\{x \in \mathbb{R}^m : |x_\alpha| < p \ , \quad \alpha = 1, \cdots, m\}$ with edge length $2p$.

Lemma 6.1 *Assume the system of conservation laws (2.1) is endowed with an involution (6.10). Fix $\hat{U} \in \mathcal{O}$ and consider the differential operator*

$$(6.16) \qquad \mathcal{L} = \sum_{\beta=1}^{m} DG_\beta(\hat{U})\partial_\beta.$$

A $2p$-periodic, L^2_{loc} function S from $\mathbb{R}^m$ to $\mathbb{R}^n$ with

$$(6.17) \qquad \int_\mathcal{K} S(x)dx = 0$$

satisfies

$$(6.18) \qquad \sum_{\alpha=1}^{m} M_\alpha \partial_\alpha S = 0$$

in the sense of distributions if and only if there is a $2p$-periodic, H^1_{loc} function χ from $\mathbb{R}^m$ to $\mathbb{R}^n$ such that

$$(6.19) \qquad S = \mathcal{L}\chi,$$

$$(6.20) \qquad \|\chi\|_{L^2(\mathcal{K})} \le ap\|S\|_{H^{-1}(\mathcal{K})} \ , \quad \|\chi\|_{H^1(\mathcal{K})} \le ap\|S\|_{L^2(\mathcal{K})} \ ,$$

where a is independent of p and S.

Proof. That (6.19) implies (6.18) follows immediately from (6.16) and (6.12).

To show necessity, expand S in Fourier series

$$(6.21) \qquad S(x) = \sum_{\xi} \exp\{\frac{i\pi}{p}(\xi \cdot x)\}X(\xi),$$

where the summation runs over all m-tuples $\xi = (\xi_1, \cdots, \xi_m)$ of integers. By (6.17), $X(0) = 0$. Note that

$$(6.22) \qquad \|S\|^2_{L^2(K)} = (2p)^m \sum_{\xi} |X(\xi)|^2 \; , \quad \|S\|^2_{H^{-1}(K)} = (2p)^m \sum_{\xi} (1 + |\xi|^2)^{-1}|X(\xi)|^2.$$

By virtue of (6.18) and (6.13), for $\xi \neq 0$,

$$(6.23) \qquad |\xi|N(|\xi|^{-1}\xi)X(\xi) = \left(\sum_{\alpha=1}^{m} \xi_\alpha M_\alpha\right) X(\xi) = 0$$

so that $X(\xi)$ lies in the kernel of $N(|\xi|^{-1}\xi)$. By assumption, the rank of $N(|\xi|^{-1}\xi)$ equals the dimension of the kernel of $\Lambda\left(|\xi|^{-1}\xi; \hat{U}\right)$. Therefore, for any $\xi \neq 0$ we may determine $Y(\xi)$ in $I\!R^n$ such that

$$(6.24) \qquad \Lambda\left(|\xi|^{-1}\xi; \hat{U}\right) Y(\xi) = -\frac{ip}{\pi}\frac{1}{|\xi|}X(\xi) ,$$

$$(6.25) \qquad |Y(\xi)| \leq \frac{ap}{2|\xi|}|X(\xi)| ,$$

for some constant a independent of p and S. It follows that the Fourier series

$$(6.26) \qquad \chi(x) = \sum_{\xi \neq 0} \exp\{\frac{i\pi}{p}(\xi \cdot x)\}Y(\xi)$$

defines a $2p$-periodic, H^1_{loc} function χ from $I\!R^m$ to $I\!R^n$, which satisfies (6.19), by virtue of (6.16), (2.2) and (6.24), as well as (6.20), by account of (6.22), (6.25) and

$$(6.27) \qquad \|\chi\|^2_{L^2(K)} = (2p)^m \sum_{\xi} |Y(\xi)|^2, \quad \|\chi\|^2_{H^1(K)} = (2p)^m \sum_{\xi} (1 + |\xi|^2)|Y(\xi)|^2.$$

This completes the proof.

Lemma 6.2 *Assume the system of conservation laws (2.1) is endowed with an involution (6.10), with involution cone C. Suppose P is a symmetric $n \times n$ matrix-valued L^∞ function on $I\!R^m$ which is uniformly positive definite in the direction of C, i.e.,*

$$(6.28) \qquad Z^T P(x)Z \geq \mu|Z|^2 \; , \quad Z \in C \; , \quad x \in I\!R^m \; ,$$

for some $\mu > 0$, and its local oscillation is less than μ, i.e.,

$$(6.29) \qquad \limsup_{\varepsilon \downarrow 0} \; \sup_{|y-x|<\varepsilon} \; |P(y) - P(x)| < \mu - 2\delta \; ,$$

for some $\delta > 0$. It W is any L^2 function from $\mathbb{R}^m$ to $\mathbb{R}^n$ which is compactly supported in the hypercube $\mathcal{K}$ and satisfies

$$\tag{6.30} \sum_{\alpha=1}^{m} M_\alpha \partial_\alpha W = Q \ ,$$

in the sense of distributions, for some Q in H^{-1}, then

$$\tag{6.31} \int_{\mathbb{R}^m} W(x)^T P(x) W(x) dx \geq \delta \|W\|_{L^2}^2 - b\|W\|_{H^{-1}}^2 - b\|Q\|_{H^{-1}}^2 \ ,$$

where b does not depend on W or Q.

Proof. Expand W and Q in Fourier series over $\mathcal{K}$:

$$\tag{6.32} W(x) = \sum_{\xi} \exp\{\frac{i\pi}{p}(\xi \cdot x)\} X(\xi) \ , \quad x \in \mathcal{K} \ ,$$

$$\tag{6.33} Q(x) = \sum_{\xi} \exp\{\frac{i\pi}{p}(\xi \cdot x)\} Y(\xi) \ , \quad x \in \mathcal{K}.$$

Note that $Y(0) = 0$ and

$$\tag{6.34} \|W\|_{L^2}^2 = (2p)^m \sum_{\xi} |X(\xi)|^2 \ , \quad \|W\|_{H^{-1}}^2 = (2p)^m \sum_{\xi} (1 + |\xi|^2)^{-1} |X(\xi)|^2,$$

$$\tag{6.35} \|Q\|_{H^{-1}}^2 = (2p)^m \sum_{\xi} (1 + |\xi|^2)^{-1} |Y(\xi)|^2$$

Furthermore, by virtue of (6.30) and (6.13), for any $\xi \neq 0$,

$$\tag{6.36} |\xi| N(|\xi|^{-1}\xi) X(\xi) = \left(\sum_{\alpha=1}^{m} \xi_\alpha M_\alpha \right) X(\xi) = Y(\xi).$$

We may thus split $X(\xi)$ into

$$\tag{6.37} X(\xi) = \Phi(\xi) + \Psi(\xi) \ , \quad \xi \neq 0 \ ,$$

where $\Phi(\xi)$ lies in the kernel of $N(|\xi|^{-1}\xi)$ while $\Psi(\xi)$ satisfies

$$\tag{6.38} |\Psi(\xi)| \leq \frac{c}{2|\xi|} |Y(\xi)|.$$

Here and below, c will stand for a generic constant, independent of W. In turn, (6.37) induces a splitting of W into

$$\tag{6.39} W(x) = X(0) + S(x) + T(x) \ , \quad x \in \mathcal{K},$$

with

$$\tag{6.40} S(x) = \sum_{\xi \neq 0} \exp\{\frac{i\pi}{p}(\xi \cdot x)\} \Phi(\xi) \ , \quad T(x) = \sum_{\xi \neq 0} \exp\{\frac{i\pi}{p}(\xi \cdot x)\} \Psi(\xi).$$

Notice that S satisfies (6.17) and (6.18) while

(6.41)
$$\|T\|_{L^2(\mathcal{K})} \leq c\|Q\|_{H^{-1}}.$$

We now cover $\overline{\mathcal{K}}$ by the union of a finite collection $\mathcal{K}_1, \cdots, \mathcal{K}_J$ of open hypercubes, centered at points $y^1, \cdots, y^J$, such that

(6.42)
$$\sup_{x \in \mathcal{K}_I} |P(x) - P(y^I)| \leq \mu - 2\delta \ , \quad I = 1, \cdots, J.$$

With the above covering we associate a partition of unity induced by C^∞ functions $\theta_1, \cdots, \theta_J$ on $\mathbb{R}^m$ such that $spt\theta_I \subset \mathcal{K}_I \cap \mathcal{K}, I = 1, \cdots, J$, and

(6.43)
$$\sum_{I=1}^{J} \theta_I^2(x) = 1 \ , \quad x \in sptW.$$

Then

(6.44)
$$\int_{\mathbb{R}^m} W(x)^T P(x) W(x) dx = \sum_{I=1}^{J} \int_{\mathcal{K}_I} \theta_I^2(x) W(x)^T P(x) W(x) dx$$

$$= \sum_{I=1}^{J} \int_{\mathcal{K}_I} \theta_I^2(x) W(x)^T P(y^I) W(x) dx + \sum_{I=1}^{J} \int_{\mathcal{K}_I} \theta_I^2(x) W(x)^T [P(x) - P(y^I)] W(x) dx.$$

By virtue of (6.42) and (6.43),

(6.45)
$$\sum_{I=1}^{J} \int_{\mathcal{K}_I} \theta_I^2(x) W(x)^T [P(x) - P(y^I)] W(x) dx \geq -(\mu - 2\delta)\|W\|_{L^2}^2.$$

Recalling Lemma 6.1, we construct the function χ which induces S through (6.19). For each $I = 1, \cdots, J$, we split $\theta_I W$ into

(6.46)
$$\theta_I W = S_I + T_I \ ,$$

where

(6.47)
$$S_I = \mathcal{L}(\theta_I \chi) \ ,$$

(6.48)
$$T_I(x) = \theta_I(x) X(0) + \theta_I(x) V(x) - \left[\sum_{\beta=1}^{m} \partial_\beta \theta_I(x) DG_\beta(\hat{U})\right] \chi(x).$$

Clearly, S_I is square integrable, has compact support in $\mathcal{K}_I$, and

(6.49)
$$\int_{\mathcal{K}_I} S_I(x) dx = 0.$$

Furthermore, by Lemma 6.1,

(6.50)
$$\sum_{\alpha=1}^{m} M_\alpha \partial_\alpha S_I = 0.$$

Consequently, S_I may be expanded in Fourier series over $\mathcal{K}_I$,

$$(6.51) \qquad S_I(x) = \sum_\xi \exp\{\frac{i\pi}{p^I}[\xi \cdot (x - y^I)]\} Z(\xi) \ , \quad x \in \mathcal{K}_I \ ,$$

with $Z(0) = 0$ and

$$(6.52) \qquad |\xi| N(|\xi|^{-1}\xi) Z(\xi) = \left(\sum_{\alpha=1}^m \xi_\alpha M_\alpha\right) Z(\xi) = 0.$$

Thus $Z(\xi)$ lies in the complexification of $\mathcal{C}$ and so, by Parseval's relation and (6.28),

$$(6.53) \qquad \int_{\mathcal{K}_I} S_I(x)^T P(y^I) S_I(x) dx = (2p^I)^m \sum_\xi Z(\xi)^* P(y^I) Z(\xi)$$

$$\geq \mu(2p^I)^m \sum_\xi |Z(\xi)|^2 = \mu \int_{\mathcal{K}_I} |S_I(x)|^2 dx.$$

Moreover, from (6.48), (6.34), (6.41), (6.20), and (6.39) we infer

$$(6.54) \qquad \int_{\mathcal{K}_I} |T_I(x)|^2 dx \leq c\|W\|_{H^{-1}}^2 + c\|Q\|_{H^{-1}}^2.$$

We now return to (6.44). From (6.46), (6.53), and (6.54) it follows that

$$(6.55) \qquad \int_{\mathcal{K}_I} \theta_I^2(x) W(x)^T P(y^I) W(x) dx$$

$$\geq (1 - \frac{\delta}{2\mu}) \int_{\mathcal{K}_I} S_I(x)^T P(y^I) S_I(x) dx - \frac{2\mu}{\delta} \int_{\mathcal{K}_I} T_I(x)^T P(y^I) T_I(x) dx$$

$$\geq (\mu - \frac{\delta}{2}) \int_{\mathcal{K}_I} |S_I(x)|^2 dx - c\|W\|_{H^{-1}}^2 - c\|Q\|_{H^{-1}}^2.$$

Again by (6.46) and (6.54),

$$(6.56) \qquad \int_{\mathcal{K}_I} |S_I(x)|^2 dx \geq (1 - \frac{\delta}{2\mu}) \int_{\mathcal{K}_I} \theta_I^2(x) |W(x)|^2 dx - \frac{2\mu}{\delta} \int_{\mathcal{K}_I} |T_I(x)|^2 dx$$

$$\geq (1 - \frac{\delta}{2\mu}) \int_{\mathcal{K}_I} \theta_I^2(x) |W(x)|^2 dx - c\|W\|_{H^{-1}}^2 - c\|Q\|_{H^{-1}}^2.$$

Combining (6.44), (6.45), (6.55), (6.56) and (6.43), we arrive at (6.31). This completes the proof.

The following proposition extends Theorem 4.1 to the situation where involutions are present and the entropy is convex only in the direction of the involution cone.

Theorem 6.1 *Assume the system of conservation laws (2.1) is endowed with an involution (6.10), with involution cone $\mathcal{C}$, and an entropy η, with $D^2\eta(U)$ positive definite in the direction of $\mathcal{C}$, uniformly on compact subsets of $\mathcal{O}$. Suppose the initial data U_0 are continuously differentiable on $\mathbb{R}^m$, take values in a compact subset of $\mathcal{O}$, are constant, say $\tilde{U}$, outside a bounded subset of $\mathbb{R}^m$, satisfy the involution on $\mathbb{R}^m$, and $\nabla U_0 \in H^\ell$ for some $\ell > m/2$. Then there exists $T_\infty, 0 < T_\infty \leq \infty$, and a unique* *continuously differentiable function U on $I\!R^m \times [0, T_\infty)$, taking values in $\mathcal{O}$, which is a classical solution of the initial-value problem (2.1), (2.3) on $[0, T_\infty)$. Furthermore,*

$$(6.57) \qquad \nabla U(\cdot, t) \in C^0([0, T_\infty); H^\ell).$$

The interval $[0, T_\infty)$ is maximal, in the sense that whenever $T_\infty < \infty$

$$(6.58) \qquad \|\nabla U(\cdot, t)\|_{L^\infty} \to \infty \, , \text{ as } t \uparrow T_\infty$$

and/or the range of $U(\cdot, t)$ escapes from every compact subset of $\mathcal{O}$ as $t \uparrow T_\infty$.

Proof. It suffices to retrace the steps of the proof of Theorem 4.1. In the definition of the metric space $\mathcal{F}$ the stipulation should be added that its members are constant, $\tilde{U}$, outside a large ball in $I\!R^m$.

The first snag we hit is that (4.15) no longer applies, since $D^2\eta(V)$ may now be positive definite only in the direction of $\mathcal{C}$. To remove this obstacle, we first note that by (4.11) and (6.12)

$$(6.59) \qquad \sum_{\beta=1}^{m} M_\beta \partial_\beta U_r = Q \, ,$$

where

$$(6.60) \qquad Q = - \sum_{\alpha,\beta=1}^{m} M_\beta \int_0^t \partial_\beta [DG_\alpha(V)] \partial_\alpha U_r d\tau$$

$$+ \sum_{\alpha,\beta=1}^{m} M_\beta \partial_\beta \int_0^t \{DG_\alpha(V)\partial^r \partial_\alpha U - \partial^r[DG_\alpha(V)\partial_\alpha U]\} d\tau.$$

·Applying Lemma 6.2, with $P = D^2\eta(V)$ and $W = U_r$, we obtain

$$(6.61) \qquad \int_{I\!R^m} U_r^T D^2\eta(V) U_r dx \geq \delta\|U_r\|_{L^2}^2 - c\|U_r\|_{H^{-1}}^2 - c\|Q\|_{H^{-1}}^2.$$

Integrating (4.11) with respect to t yields

$$(6.62) \qquad \|U_r(\cdot, t)\|_{H^{-1}} \leq c\|\nabla U_0(\cdot)\|_\ell + c\omega \int_0^t \|\nabla U(\cdot, \tau)\|_\ell d\tau.$$

Furthermore, (6.60) together with (4.6) and (4.12) imply

$$(6.63) \qquad \|Q(\cdot, t)\|_{H^{-1}} \leq c\omega \int_0^t \|\nabla U(\cdot, \tau)\|_\ell d\tau.$$

By employing (6.61), (6.62), (6.63) as a substitute for (4.15), we establish, in the place of (4.16), the new estimate

$$(6.64) \qquad \|\nabla U(\cdot, t)\|_\ell^2 \leq c\|\nabla U_0(\cdot)\|_\ell^2 + c\omega(1 + \omega T) \int_0^t \|\nabla U(\cdot, \tau)\|_\ell^2 d\tau$$

whence we deduce that, when ω is sufficiently large and T is sufficiently small, $\sup_{[0,T]} \|\nabla U(\cdot, t)\|_\ell \leq \omega$, as required for the proof.

A similar procedure is used to compensate for the failure of (4.21): We use (4.18) and (6.12) to get

$$(6.65) \qquad \sum_{\beta=1}^{m} M_\beta \partial_\beta (U - \overline{U}) = Q \ ,$$

where

$$(6.66) \qquad Q = - \sum_{\alpha,\beta=1}^{m} M_\beta \int_0^t \partial_\beta [DG_\alpha(V)] \partial_\alpha (U - \overline{U}) d\tau$$

$$- \sum_{\alpha,\beta=1}^{m} M_\beta \partial_\beta \int_0^t [DG_\alpha(V) - DG_\alpha(\overline{V})] \partial_\alpha \overline{U} d\tau \ ,$$

and then apply Lemma 6.2, with $P = D^2\eta(V), W = U - \overline{U}$, to get

$$(6.67) \qquad \int_{\mathbb{R}^m} (U - \overline{U})^T D^2\eta(V)(U - \overline{U}) dx \geq \delta \|U - \overline{U}\|_{L^2}^2 - c\|U - \overline{U}\|_{H^{-1}}^2 - c\|Q\|_{H^{-1}}^2 .$$

Integrating (4.18) with respect to t, we obtain the estimate

$$(6.68) \qquad \|(U - \overline{U})(\cdot, t)\|_{H^{-1}} \leq c\omega \int_0^t \{\|(U - \overline{U})(\cdot, \tau)\|_{L^2} + \|(V - \overline{V})(\cdot, \tau)\|_{L^2}\} d\tau.$$

Moreover, (6.66) together with (4.6) imply

$$(6.69) \qquad \|Q(\cdot, t)\|_{H^{-1}} \leq c\omega \int_0^t \{\|(U - \overline{U})(\cdot, \tau)\|_{L^2} + \|(V - \overline{V})(\cdot, \tau)\|_{L^2}\} d\tau.$$

We employ (6.67), (6.68) and (6.69) as a substitute for (4.21). This yields, in the place of (4.22), the new estimate

$$(6.70) \qquad \|(U - \overline{U})(\cdot, t)\|_{L^2}^2 \leq c\omega(1 + \omega T) \int_0^t \{\|(U - \overline{U})(\cdot, \tau)\|_{L^2}^2 + \|(V - \overline{V})(\cdot, \tau)\|_{L^2}^2\} d\tau.$$

From (4.9), (6.70) and Gronwall's inequality we deduce

$$(6.71) \qquad \rho(U, \overline{U}) \leq [c\omega T(1 + \omega T)]^{1/2} \exp[c\omega T(1 + \omega T)] \rho(V, \overline{V})$$

which verifies that, for T small, the map that carries $V \in \mathcal{F}$ to the solution $U \in \mathcal{F}$ of (4.10), (2.3) is a contraction.

Apart from the above modifications, the proofs of Theorems 6.1 and 4.1 are identical.

To illustrate the use of Theorem 6.1, we apply it to the system (6.1), with involution (6.7). A simple calculation shows that the involution cone $\mathcal{C}$ consists of all vectors in $\mathbb{R}^{m^2+m}$ of the form $(\xi \otimes \nu, w)$, with arbitrary ξ, ν and w in $\mathbb{R}^m$. Thus, the entropy $\eta = \varepsilon(F) + \frac{1}{2}|v|^2$ is convex in the direction of $\mathcal{C}$ provided that $\varepsilon(F)$ satisfies (6.3). Consequently, Theorem 6.1 establishes local existence of classical solutions for the system of balance laws of isentropic, adiabatic thermoelasticity under the physically natural assumption that the internal energy is a rank-one convex function of the deformation gradient.

Theorem 5.1 may also be similarly extended to the situation where the entropy is convex just in the direction of the involution cone:

Theorem 6.2 *Assume the system of conservation laws (2.1) is endowed with an involution (6.10), with involution cone $\mathcal{C}$, and with a proper entropy η with $D^2\eta(U)$ positive definite in the direction of $\mathcal{C}$, uniformly on compact subsets of $\mathcal{O}$. Suppose $\overline{U}$ is a classical solution of (2.1) on a bounded time interval $[0, T)$, taking values in a convex, compact subset $\mathcal{D}$ of $\mathcal{O}$, with initial data $\overline{U}_0$ that satisfy the involution. Let U be any admissible weak solution of (2.1) on $[0, T)$, taking values in $\mathcal{D}$, which coincides with $\overline{U}$ outside some ball in $\mathbb{R}^m$, has local oscillation*

$$(6.72) \qquad \limsup_{\varepsilon\downarrow 0} \ \sup_{|y-x|<\varepsilon} |U(y,t) - U(x,t)| < \kappa \ , \quad 0 \le t < T \ ,$$

and initial data U_0 satisfying the involution. If κ is small, then

$$(6.73) \qquad \int_{\mathbb{R}^m} |U(x,t) - \overline{U}(x,t)|^2 dx \le a \int_{\mathbb{R}^m} |U_0(x) - \overline{U}_0(x)|^2 dx$$

holds for $t \in [0, T)$, and some constant a that depends on $\mathcal{D}$, on T and on the Lipschitz constant of $\overline{U}$. In particular, $\overline{U}$ is the unique admissible weak solution of (2.1) with values in $\mathcal{D}$, small local oscillation and initial data $\overline{U}_0$.

Proof. Retracing the steps of the proof of Theorem 5.1, we rederive (5.13). Letting $R \uparrow \infty$ and $\varepsilon \downarrow 0$ in (5.13), taking into account that $U - \overline{U}$ vanishes outside some ball, $\overline{U}$ is Lipschitz, Theorem 4.1.1, and that η is proper, we arrive again at (5.14), with $R = \infty$:

$$(6.74) \qquad \int_{\mathbb{R}^m} h(U(x,t), \overline{U}(x,t)) dx \le \int_{\mathbb{R}^m} h(U_0(x), \overline{U}_0(x)) dx$$

$$- \int_0^t \int_{\mathbb{R}^m} \sum_{\alpha=1}^m \partial_\alpha \overline{U}^T D^2\eta(\overline{U}) Z_\alpha(U, \overline{U}) dx d\tau.$$

From (5.2),

$$(6.75) \qquad h(U, \overline{U}) = (U - \overline{U})^T P(U, \overline{U})(U - \overline{U}),$$

where

$$(6.76) \qquad P(U, \overline{U}) = \int_0^1 \int_0^w D^2\eta(\overline{U} + z(U - \overline{U})) dz dw.$$

In particular,

$$(6.77) \qquad Z^T P(U, \overline{U}) Z \ge \mu |Z|^2 \ , \quad Z \in \mathcal{C} \ ,$$

for some $\mu > 0$. Therefore, when κ in (6.72) is so small that the local oscillation of $P(U(x,t), \overline{U}(x,t))$ is less than μ, we may apply Lemma 6.2, with $W = U - \overline{U}$ and $Q = 0$, to get

$$(6.78) \qquad \int_{\mathbb{R}^m} h(U(x,t), \overline{U}(x,t)) dx \ge \delta \|U(\cdot, t) - \overline{U}(\cdot, t)\|_{L^2}^2 - c\|U(\cdot, t) - \overline{U}(\cdot, t)\|_{H^{-1}}^2$$

for some $\delta > 0$. We estimate the second term on the right-hand side of (6.78) as follows:

$$(6.79) \qquad \|U(\cdot, t) - \overline{U}(\cdot, t)\|_{H^{-1}} \le \|U_0(\cdot) - \overline{U}_0(\cdot)\|_{H^{-1}}$$

$$+ \int_0^t \|\partial_t \{U(\cdot,\tau) - \overline{U}(\cdot,\tau)\}\|_{H^{-1}} d\tau \ ,$$

(6.80) $\qquad \|\partial_t\{U(\cdot,\tau) - \overline{U}(\cdot,\tau)\}\|_{H^{-1}} = \| \sum_{\alpha=1}^m \partial_\alpha \{G_\alpha(U(\cdot,\tau)) - G_\alpha(\overline{U}(\cdot,\tau))\}\|_{H^{-1}}$

$$\leq \sum_{\alpha=1}^m \|G_\alpha(U(\cdot,\tau)) - G_\alpha(\overline{U}(\cdot,\tau))\|_{L^2} \leq c\|U(\cdot,\tau) - \overline{U}(\cdot,\tau)\|_{L^2}.$$

Furthermore, by (5.4), $Z_\alpha(U,\overline{U})$ is of quadratic order in $U - \overline{U}$. Therefore, combining (6.78), (6.79) and (6.80), we deduce from (6.74):

(6.81) $\qquad \|U(\cdot,t) - \overline{U}(\cdot,t)\|_{L^2}^2 \leq c\|U_0(\cdot) - \overline{U}_0(\cdot)\|_{L^2}^2$

$$+c \int_0^t \|U(\cdot,\tau) - \overline{U}(\cdot,\tau)\|_{L^2}^2 d\tau + c\{\int_0^t \|U(\cdot,\tau) - \overline{U}(\cdot,\tau)\|_{L^2} d\tau\}^2$$

whence (6.73) follows. This completes the proof.

Remark 6.1 The condition that the entropy is proper has to be imposed as an explicit assumption in Theorem 6.2 whereas in Theorem 5.1 it is a byproduct of the assumption that the entropy is convex. Note, however, that if the map $t \mapsto U(\cdot,t)$ is strongly continuous on $L_{loc}^1(\mathbb{R}^m)$, for instance when U is in BV_{loc}, then it is not necessary to assume that the entropy is proper.

The natural extension of the notion of convexity, in the presence of involutions, is provided by the following (compare with [Mo,D]):

Definition 6.3 An entropy η for the system of conservation laws (2.1), endowed with an involution (6.10), is called *quasiconvex* if for any constant vector $\hat{U} \in \mathcal{O}$, any hypercube $\mathcal{K}$ in $\mathbb{R}^m$ and any Lipschitz function χ from $\mathcal{K}$ to $\mathbb{R}^n$, with compact support in $\mathcal{K}$,

(6.82) $$\int_{\mathcal{K}} \eta(\hat{U}) dx \leq \int_{\mathcal{K}} \eta(\hat{U} + \mathcal{L}\chi) dx \ ,$$

where $\mathcal{L}$ is the differential operator defined by (6.16).

The relevance of quasiconvexity is demonstrated by the following proposition, whose proof may be found in [D].

Theorem 6.3 *Assume the system of conservation laws* (2.1) *is endowed with an involution* (6.10), *with involution cone $\mathcal{C}$. Then any quasiconvex entropy is necessarily proper and convex in the direction of $\mathcal{C}$.*

A partial converse of the above statement is also valid: If all the constraints on solutions of (2.1) are induced by the involution (6.10), then any proper entropy is necessarily quasiconvex.

Roughly, quasiconvexity stipulates that the uniform state minimizes the total entropy, among all states that are compatible with the involution and have the same total "mass". This is in the spirit of classical thermostatics, recalling that what we call here "entropy" in the applications is actually minus the physical entropy.

Any convex entropy is quasiconvex, by virtue of Jensen's inequality. However, whereas convexity may be tested easily, by calculation, quasiconvexity, as defined by Definition 6.3, is an implicit condition that is hard to verify in practice. In view of Theorem 6.3, it is tempting to conjecture that any entropy which is convex in the direction of the involution cone is necessarily quasiconvex. This is indeed the case when the entropy is quadratic: $\eta = U^T A U$. In general, however, quasiconvexity is a stricter condition than mere convexity in the direction of the involution cone.

Let us illustrate the above in the framework of our prototypical example, the system of balance laws (6.1) of isentropic, adiabatic thermoelasticity, with involution (6.7) and entropy (6.4). In that case η is quasiconvex when $\varepsilon(F)$ is quasiconvex in the sense of Morrey [Mo]: For any constant deformation gradient $\hat{F}$, any hypercube $\mathcal{K}$ in $I\!\!R^m$ and any Lipschitz function χ from $\mathcal{K}$ to $I\!\!R^m$, with compact support in $\mathcal{K}$,

$$(6.83) \qquad \int_{\mathcal{K}} \varepsilon(\hat{F}) dx \leq \int_{\mathcal{K}} \varepsilon(\hat{F} + \nabla\chi) dx.$$

In other words, a homogeneous deformaation of $\mathcal{K}$ minimizes the total internal energy among all placements of $\mathcal{K}$ with the same boundary values. Any quasiconvex internal energy is rank-one convex (6.3), i.e., η is convex in the direction of the convolution cone. On the other hand, examples have been constructed of $\varepsilon(F)$ that are rank-one convex but not quasiconvex [S]. A sufficient condition for $\varepsilon(F)$ to be quasiconvex is that it be *polyconvex*, in the sense of Ball [B], i.e., in the physically relevant case $m = 3$:

$$(6.84) \qquad \varepsilon(F) = \sigma(F, F^*, \det F) \ ,$$

where F^* denotes the matrix of cofactors of F, $F^* = (\det F) F^{-1}$, and σ is a convex function on $I\!\!R^{19}$. When the elastic material is isotropic, $\varepsilon(F) = \tilde{\varepsilon}(I_1, \cdots, I_m)$, where $(I_1, \cdots, I_m)$ are the principal invariants of $F^T F$ (see [TN]). If $\tilde{\varepsilon}$ is convex, then $\hat{\varepsilon}(F)$ is polyconvex [B].

References

[B] J.M. Ball, Convexity conditions and existence theorems in nonlinear elasticity. Arch. Rat. Mech. Anal. **63** (1977), 337-403.

[B1] G. Boillat, Sur l' existence et la recherche d' équations de conservation supplémentaires pour les systèmes hyperboliques. C.R. Acad. Sc. Paris **278** (1974), 909.

[B2] G. Boillat, Involutions des systèmes conservatifs. C.R. Acad. Sc. Paris **307** (1988), 891-894.

[D] B. Dacorogna, *Weak Continuity and Weak Lower Semicontinuity of Non-Linear Functionals.* Lecture Notes in Math. No. 922. Berlin: Springer-Verlag 1982.

[Da] C.M. Dafermos, Quasilinear hyperbolic systems with involutions. Arch. Rat. Mech. Anal. **94** (1986), 373-389.

[Di] R.J. DiPerna, Uniqueness of solutions to hyperbolic conservation laws. Indiana U. Math. J. **28** (1979), 137-188.

[FL] K.O. Friedrichs and P.D. Lax, Systems of conservation equations with a convex extension. Proc. Nat. Acad. Sci. USA **68** (1971), 1686-1688.

[G] S.K. Godunov, An interesting class of quasilinear systems. Sov. Math. **2** (1961), 947-948.

[M] A. Majda, *Compressible Fluid Flow and Systems of Conservation Laws in Several Space Variables.* New York: Springer-Verlag 1984.

[Mo] C.B. Morrey, Quasiconvexity and the lower semicontinuity of multiple integrals, Pacific J. Math. **2** (1952), 25-53.

[RS] T. Ruggeri and A. Strumia, Main field and convex covariant density for quasilinear hyperbolic systems. Ann. Inst. Henri Poincaré. **34** (1981), 65-84.

[S] V. Šverák, Rank-one convexity does not imply quasiconvexity. *Proc. Roy. Soc. Edinburgh* **120** (1992), 185-189.

[TN] C. Truesdell, and W. Noll, *The Nonlinear Field Theories of Mechanics. Handbuch der Physik III/3*, ed. S. Flügge. Berlin: Springer-Verlag 1965.

Outline of a theory
of the KdV equation

by Peter D. Lax

Introduction

These lectures are meant to be an introduction to the fascinating theory of the
Korteweg-de Vries (KdV) equation for those who are unfamiliar with this topic.

Chapter 0 is a telegrammatic review of classical Hamiltonian theory, including a
thumbnail sketch of completely integrable Hamiltonian systems. Chapter I describes
the neoclassical results for the KdV equation, the basic discoveries of Kruskal and
Zabusky, Gardner, Lax, Faddeev-Zakharov, and others, although not in the order
of their original discovery. Chapter II is a straightforward account of direct and
inverse scattering theory for the one dimensional Schroedinger equation. Chapter III
presents the Gardner-Greene-Kruskal-Miura solution of the initial value problem using
scattering theory.

The classical examples of completely integrable Hamiltonian systems are geodesic
flow on ellipsoids, integrated by Jacobi, Carl Neumann's study of the motion of a
particle under linear force confined to move on a sphere, and Sonya Kowalewski's integration of the motion of the top subject to a condition bearing her name. Thereafter
interest in such systems receded, until the advent of the Bohr-Sommerfeld method of
quantization. With the creation of quantum mechanics complete integrability re-
treated to the seeming wastebasket of history, only to reemerge in the late sixties
of this century, thanks to the unexpected discovery of a whole slew of new, completely integrable systems. Some of these were finite dimensional, like the finite or
periodic Toda chain, the Kac-van Moerbeke chain, the inverse cube force acting on
particles on a line, and others. Equally remarkable was the discovery of infinite dimensional, even continuum, systems, of which the first was the KdV equation, followed
by cubic Schroedinger, sine-Gordon, Benjamin-Ono, Davey-Stewartson, Kadomtsev-
Petviashvili, etc. It is with the first of these that the present series of lectures is
concerned.

I would like to emphasize that the purpose of these notes is not to take the reader
up to the leading edge of research in the rapidly growing area of completely integrable
systems, but to introduce the novice to the subject. I hope that I have achieved this
modest but worthwhile goal.

Chapter 0.   Classical Hamiltonian Mechanics

Phase space $(p_s, q_s) = (p, q)$.

Hamiltonian $H(p, q)$ a smooth, real valued function.

Hamiltonian flow: $q_t = H_p$, $p_t = -H_q$, where subscript is $\dfrac{\partial}{\partial t}$. Set $\binom{q}{p} = u$:

$$u_t = JH_u, \qquad J = \begin{pmatrix} 0 & I \\ -I & 0 \end{pmatrix}.$$

Liouville's Theorem: *H-flow is volume preserving.*

A solution of Hamilton's equation is called a *trajectory*, or *orbit*. We denote by $w(u, t)$ the orbit whose state at $t = 0$ is u : $w(u, 0) = 0$. Since H is assumed independent of t, the w form a group:

$$w(w(u, s), t) = w(u, s + t).$$

Let $F = F(q, p)$ be any smooth function in phase space. Denote differentiation along H orbits by $\dfrac{d}{dt}$. Then

$$\begin{aligned}
\frac{d}{dt} F &= F_q q_t + F_p p_t = F_q H_p - F_p H_q \\
&= F_u \cdot JH_u \overset{\text{def}}{=} \{F, H\}
\end{aligned}$$

The *Poisson bracket* $\{\ ,\ \}$ has these properties:

i) Antisymmetry: $\{F, H\} = -\{H, F\}$.

ii) Bilinearity

iii) Jacobi identity:

$$\{\{F, G\}, H\} + \{\{H, F\}, G\} + \{\{G, H\}, F\} = 0.$$

Clearly, F is constant along all H-trajectories iff $\{F, H\} = 0$; in particular, H is constant.

Theorem: *Let H and F be a pair of Hamiltonians; H-flow and F-flow commute iff $\{F, H\} = 0$.*

Proof: With each H-flow we associate the *Koopman group* of operators $\mathcal{U}_H(t)$, t in $\mathbb{R}$, acting on L^2 functions G in phase space, defined as follows:

$$(\mathcal{U}_H(t)G)(u) = G(w_H(u, t)).$$

By Liouville's theorem, each $\mathcal{U}_H(t)$ is unitary, and by the group property of the flow $w_H(u, t)$,

$$\mathcal{U}_H(t + s) = \mathcal{U}_H(t)\mathcal{U}_H(s).$$

Denote by $\mathcal{A}_H$ the infinitesimal generator of the Koopman group:

$$\frac{d}{dt}\mathcal{U}_H(t)G\Big|_{t=0} = \{G, H\} = \mathcal{A}_H G,$$

$\mathcal{A}_H$ is an antiselfadjoint first order operator. By the Jacobi identity

$$[\mathcal{A}_F, \mathcal{A}_H]G = \mathcal{A}_F \mathcal{A}_H G - \mathcal{A}_H \mathcal{A}_F G =$$
$$= \{\{G, H\}, F\} - \{\{G, F\}, H\} = \{\{H, F\}, G\}.$$

This can be expressed as

$$[\mathcal{A}_F, \mathcal{A}_H] = \mathcal{A}_{\{H,F\}}.$$

If $\{H, F\} = 0$, $\mathcal{A}_F$ and $\mathcal{A}_H$ commute. It follows that then $\mathcal{U}_H(t)$ and $\mathcal{U}_F(s)$ commute also, and therefore so do the flows $w_H(u,t)$ and $w_F(u,s)$.

If $\{F, H\} = 0$, F and H are said to be *in involution*.

Canonical transformations

Let $v = v(u)$ be a transformation of phase space. Denote its Jacobian matrix by M:

$$\frac{\partial v}{\partial u} = M, \qquad \frac{\partial v_i}{\partial u_j} = M_{ij}.$$

Suppose u satisfies a Hamiltonian equation $u_t = JH_u$. Then v satisfies

$$v_t = \frac{\partial v}{\partial u}u_t = Mu_t = MJH_u = MJM^T K_v,$$

where $K(v(u)) = H(u)$. This is of Hamiltonian form iff $MJM^T = J$. Such a matrix M is called a *symplectic matrix*. A transformation whose Jacobian is symplectic is called canonical; it preserves the Hamiltonian form of equations.

Symplectic matrices form a group; therefore so do canonical transformations.

Completely integrable systems

Let $F_1, \ldots, F_n$ be functions defined on $2n$-dimensional phase space, with the following properties:

i) F_i, F_j are in involution: $\{F_i, F_j\} = 0$.

ii) The F_j are independent, i.e. their gradients are linearly independent except on $n - 1$ dimensional submanifolds.

Such a collection of Hamiltonians is called *completely integrable*.

Let z_0 be a point of phase space where grad F_j are linearly independent. The intersection of the level sets

$$F_i(z) = F_i(z_0)$$

is an n-dimensional submanifold of phase space; denote is as $M(z_0)$.

Theorem: $M(z_0)$ *is an invariant submanifold under all F_j-flows.*

Liouville's Theorem: *Given a completely integrable collection of Hamiltonians, there is a canonical change of variables to $P_1, \ldots, P_n$, $Q_1, \ldots, Q_n$ such that the P_i are functions of $F_1, \ldots, F_n$. The P_i are called* action variables, *the Q_i angle variables.*

Take H to be one of the F_j; since $F_j = F_j(P)$ is independent of Q, it follows from $P_t = -H_Q$ that the action variables P are constant along trajectories. It follows from $Q_t = H_p$ that the angle variables Q are linear along trajectories. When the phase space is compact, these trajectories lie on tori.

For proof, see Arnold's Mechanics. The construction of P and Q can be accomplished *without* solving differential equations.

Chapter I. The KdV equation as a Hamiltonian system

Phase space is the space of L^2 *functions* u on $\mathbb{R}$, or S^1. The scalar product is denoted as (,). Thus functions H, F etc. on phase space are nonlinear *functionals* on L^2. Most of these functionals are unbounded.

The *gradient* G_F of F at u is a *function* that satisfies

$$\frac{d}{d\epsilon}F(u + \epsilon v)\Big|_{\epsilon=0} = (v, G_F) \qquad (1)$$

for all variations v.

We define the *second derivative* of F at u as the *linear operator* S_F that satisfies

$$\frac{d}{d\eta}G_F(u + \eta w)\Big|_{\eta=0} = S_F w \qquad (2)$$

for all variations w. Directly in terms of F

$$\frac{d^2}{d\eta d\epsilon}F(u + \epsilon v + \eta w)\Big|_{\epsilon=0,\eta=0} = (v, S_F w). \qquad (2)'$$

This shows that S_F is a *symmetric* operator.

Gardner and Faddeev-Zakharov define a Hamiltonian structure by the Poisson bracket

$$\{F, H\} = (G_F, \partial G_H); \qquad (3)$$

$\partial = \dfrac{d}{dx}$ is differentiation on $\mathbb{R}$ or S^1. The corresponding Hamiltonian flow is

$$u_t = \partial G_H. \qquad (3)'$$

Clearly, (3) is bilinear. Since ∂ is anti-symmetric, so is (3). We compute now the gradient of $\{F, H\}$:

$$\begin{aligned}
\frac{d}{d\epsilon}\{F, H\}(u + \epsilon v) &= \frac{d}{d\epsilon}(G_F(u + \epsilon v), \partial G_H(u + \epsilon v)) \\
&= (S_F v, \partial G_H) + (G_F, \partial S_H v) = \\
&= (v, S_F \partial G_H - S_H \partial G_F);
\end{aligned} \qquad (5)$$

in the last step we have used the symmetry of S and the antisymmetry of ∂. So

$$G_{\{F,H\}} = S_F \partial G_H - S_H \partial G_F. \qquad (5)'$$

The *Jacobi identity*, using (5)',

$$\begin{aligned}
&\{\{F, H\}, K\} + \{\{K, F\}, H\} + \{\{H, K\}, F\} = \\
&(G_{\{F,H\}}, \partial G_K) + (G_{\{K,F\}}, \partial G_H) + (G_{\{H,K\}}, \partial G_F) = \\
&(S_F \partial G_H - S_H \partial G_F, \partial G_K) + (S_K \partial G_F - S_F \partial G_K, \partial G_H) + \\
&(S_H \partial G_K - S_K \partial G_H, \partial G_F) = 0
\end{aligned} \qquad (6)$$

follows from the symmetry of $\mathcal{S}_F, \mathcal{S}_H, \mathcal{S}_K$.

Choose as Hamiltonian

$$H^{(1)}(u) = \int \left(\frac{1}{2} u_x^2 - \frac{1}{6} u^3 \right) dx. \tag{7}$$

Its gradient is

$$G_{H^{(1)}} = -u_{xx} - \frac{1}{2} u^2. \tag{7}'$$

Setting this into (3)' yields the celebrated KdV equation

$$u_t + u u_x + u_{xxx} = 0. \tag{8}$$

There is another possible Hamiltonian structure, employing the Lenard operator

$$\mathcal{N} = \partial^3 + \frac{2}{3} u \partial + \frac{1}{3} u_x, \tag{9}$$

whose coefficients depend on u. For any choice of u, $\mathcal{N}$ is linear and antisymmetric. We define the new Poisson bracket as

$$\{F, H\}_{\text{new}} = (G_F, \mathcal{N} \partial G_H). \tag{10}$$

The corresponding Hamiltonian flow is

$$u_t = \mathcal{N} G_H. \tag{10}'$$

Clearly (10) is bilinear. Since $\mathcal{N}$ is antisymmetric, so is (10). We compute now the gradient of $\{F, H\}_{\text{new}}$. Since $\mathcal{N}$ depends on u, there is an extra term in addition to the analogous ones appearing in (5). This term is

$$\left(G_F, \left(\frac{2}{3} v \partial + \frac{1}{3} v_x \right) G_H \right) =$$
$$\frac{2}{3} (G_F \partial G_H, v) - \frac{1}{3} (\partial (G_H G_F), v) =$$
$$\frac{1}{3} (G_F \partial G_H - G_H \partial G_F, v).$$

So

$$\begin{aligned} G_{\{F,H\}_{\text{new}}} &= (\mathcal{S}_F \mathcal{N} G_H - \mathcal{S}_H \mathcal{N} G_F) + \\ &\quad \frac{1}{3} (G_F \partial G_H - G_H \partial G_F). \end{aligned} \tag{11}$$

The verification of the Jacobi identity is now easy. For the first group of terms it goes the same way as in (6); for the second group it goes even easier.

We choose now as Hamiltonian

$$H^{(0)} = -\int \frac{1}{2} u^2 \, dx; \tag{12}$$

its gradient is

$$G_{H^{(0)}} = -u. \tag{12}'$$

Setting this into the Hamilton equation (10)' yields *again* the KdV equation (8)!

The Hamiltonian is constant along every H-trajectory. Therefore both $H^{(0)}$ and $H^{(1)}$ are constant along every KdV flow, i.e. they are *conserved quantities*. There are infinitely many more; they can be constructed by observing that since (3)' with $H^{(1)}$ and (10)' with $H^{(0)}$ both are the KdV equation,

$$\partial G_{H^{(1)}} = \mathcal{N} G_{H^{(0)}}. \tag{13}$$

Theorem 1: *There is an infinity sequence of Hamiltonians $H^{(k)}$ satisfying the recursion*

$$\partial G_{k+1} = \mathcal{N} G_k \tag{13$_k$}$$

where G_k denotes $G_{H^{(k)}}$. Each $H^{(k)}$ is of the form

$$\int Q_k \, dx,$$

Q_k a polynomial in $u, \partial u, \ldots, \partial^k u$; $H^{(0)}$ and $H^{(1)}$ are defined by (12) and (7).

Proof: We construct recursively the gradients of the $H^{(k)}$. We start with the following elementary calculus result:

Lemma 2: *Let P be a polynomial in u and its derivatives of order up to j, such that*

$$\int P(u) \, d = 0 \tag{14}$$

for all C^j functions u with compact support. Then P is of the form

$$P = \partial G, \tag{14'}$$

where G is a polynomial in u and its derivatives up to order $j - 1$.

Proof: Clearly every P of the form (14)' satisfies (14). Conversely suppose that (14) holds for all u in C_0^j. Take any value x and define the function w by

$$w(y) = \begin{cases} u(y) & \text{for } y \leq x \\ \sum_0^{j-1} m_i \varphi_i(y - x) & \text{for } x \leq y \end{cases} \tag{15}$$

where $\varphi_i(y)$ are smooth functions of compact support, satisfying

$$\partial^k \varphi_i \big|_{y=0} = \delta_{ik} \tag{15'}$$

and

$$m_i = \partial^i u \big|_{y=x}. \tag{15''}$$

Applying (14) to w defined in (15) shows that

$$\int\limits_{-\infty}^{x} P(u) \, dy = - \int\limits_{x}^{\infty} P(w) \, dy.$$

The right hand side depends only on $\partial^i u \big|_{y=x}$, $i = 0, \ldots, j - 1$ and is a polynomial in these variables, as claimed in the Lemma. $\qquad \square$

We construct the functionals $H^{(k)}$ recursively. It is convenient to start at $k = -1$, with

$$H^{(-1)} = -\int 3u\,dx, \qquad G_{-1} = -3. \tag{16}$$

Formula (12)' for G_0 and (9) for $\mathcal{N}$ shows that the recursion relation $(13)_k$ is satisfied for $k = -1$, as it is for $k = 0$.

Suppose G_k is known; we determine G_{k+1} from equation $(13)_k$. According to our Lemma 2, such a G_{k+1} exists iff

$$\int \mathcal{N}G_k\,dx = 0 \tag{17}$$

for all smooth u of compact support.

In view of the definition (16) of G_{-1}, we can rewrite (17) as

$$(\mathcal{N}G_k, G_{-1}) = 0. \tag{18}$$

Using the antisymmetry of $\mathcal{N}$ and ∂, and the induction hypothesis that $\partial G_{j+1} = \mathcal{N}G_j$ for $j < k$ we get

$$
\begin{aligned}
(\mathcal{N}G_k, G_{-1}) &= -(G_k, \mathcal{N}G_{-1}) = -(G_k, \partial G_0) \\
&= (\partial G_k, G_0) = (\mathcal{N}G_{k-1}, G_0) = \ldots \\
&= \begin{cases} (\partial G_{k/2}, G_{k/2}) & \text{if } k \text{ is even} \\ (\mathcal{N}G_{(k-1)/2}, G_{(k-1)/2}) & \text{if } k \text{ is odd.} \end{cases}
\end{aligned}
\tag{19}
$$

Because of antisymmetry either of these scalar products is zero, proving (18).

Having determined G_{k+1}, we prove that it is a gradient. According to the *infinite dimensional Poincaré lemma*, it suffices to verify that the gradient of G_{k+1} defined by (2) and denoted here as S_{k+1}, is a *symmetric* operator.

To see this we take the derivative of $(13)_k$ with respect to u in the direction v:

$$\partial S_{k+1}v = \left(\frac{2}{3}v\partial + \frac{1}{3}v'\right)G_k + \mathcal{N}S_k v; \tag{19_k}$$

here $'$ denotes differentiation with respect to x, and we have used the definition (9) of $\mathcal{N}$. We rewrite $(19)_k$ in operator form as

$$\partial S_{k+1} = \frac{1}{3}G_k\partial + \frac{2}{3}G_k' + \mathcal{N}S_k. \tag{$19_k'$}$$

Similarly

$$\partial S_k = \frac{1}{3}G_{k-1}\partial + \frac{2}{3}G_{k-1}' + \mathcal{N}S_{k-1}. \tag{$19_{k-1}'$}$$

Multiply $(19)_k'$ by ∂ on the right, multiply $(19)_{k-1}'$ by $\mathcal{N}$ on the right and subtract:

$$
\begin{aligned}
\partial S_{k+1}\partial &= -\mathcal{N}S_{k-1}\mathcal{N} + \mathcal{N}S_k\partial + \partial S_k\mathcal{N} \\
&\quad + \frac{1}{3}G_k\partial^2 + \frac{2}{3}G_k'\partial - \frac{1}{3}G_{k-1}\partial\mathcal{N} - \frac{2}{3}G_{k-1}'\mathcal{N}.
\end{aligned}
\tag{20}
$$

By induction hypothesis S_{k-1} and S_k are symmetric operators. Since $\mathcal{N}$ and ∂ are antisymmetric, the first term on the right in (20) is symmetric, and so is the sum of the next two terms.

The remaining four operator are of fourth order; that their sum is symmetric can be verified by a calculation, using the identity $\partial G_k = \mathcal{N} G_{k-1}$.

This shows that $\partial S_{k+1} \partial$ is symmetric; the symmetry of S_{k+1} follows easily. So G_{k+1} is the gradient of $H^{(k+1)}$. □

Remark: It was shown in Lax [12] that functionals $H^{(k)}$ can be constructed for all *negative* indices so that $(13)_k$ is satisfied.

Corollary 3: *The Hamiltonians $H^{(k)}$ are in involution with respect to either Poisson structure: for all k and ℓ,*

$$(G_k, \partial G_\ell) = 0, \qquad (G_k, \mathcal{N} G_\ell) = 0.$$

Proof: This follows from the string of identities $(18)'$. □

It follows from the corollary that each $H^{(\ell)}$ is a conserved quantity for either of the Hamiltonian flows with $H = H^{(k)}$, and that all these Hamiltonian flows commute.

We call the sequence of Hamiltonian flows

$$u_t + \partial G_k = 0 \tag{21}$$

the *hierarchy* of KdV flows.

The above set of conserved functional $H^{(k)}$ were first constructed, in another way, by Kruskal, Zabusky, Miura and Gardner. They also discovered another set of conserved quantities associated with the Schroedinger operator

$$\mathcal{L} = -\partial^2 - u/2. \tag{22}$$

Here the potential u depends on a parameter t.

We appeal now to the following general principle

Theorem 4 (Lax) *Let $\mathcal{L}(t)$ be a one-parameter family of selfadjoint operators with the same domain that depends differentiably on t. Let $\mathcal{B}(t)$ be a one-parameter family of antiselfadjoint operators with common domain, depending continuously on t. Suppose $\mathcal{L}$ and $\mathcal{B}$ satisfy*

$$\mathcal{L}_t = \mathcal{B}\mathcal{L} - \mathcal{L}\mathcal{B}. \tag{23}$$

Then the operators $\mathcal{L}(t)$ are unitarily equivalent; in particular, their eigenvalues are independent of t.

Proof: The operator function $\mathcal{B}(t)$ generates a family of unitary operators $\mathcal{U}(t)$ via the differential equations

$$\mathcal{U}_t = \mathcal{B}\mathcal{U}, \qquad \mathcal{U}(0) = \mathcal{I}. \tag{23'}$$

It follows from (23) and $(23)'$ that

$$\frac{d}{dt}\mathcal{U}^*\mathcal{L}\mathcal{U} = 0;$$

so $\mathcal{U}^*(t)\mathcal{L}(t)\mathcal{U}(t) = \mathcal{L}(0)$, as asserted. □

For the Schroedinger operator (22) we take $\mathcal{B}$ to be a differential operator of odd order $2m + 1$:

$$\mathcal{B}_m = \partial^{2m+1} + \sum_0^{m-1} b_j \partial^{2j+1} + \partial^{2j+1} b_j,$$

chosen to be antisymmetric. The commutator of $\mathcal{B}_m$ and $\mathcal{L}$ is a symmetric operator of order $2m$; we choose the coefficients b_j so that $[\mathcal{B}_m, \mathcal{L}]$ is of order zero. Because of symmetry, this imposes m conditions. In particular

$$m = 0, \qquad \mathcal{B}_0 = \partial, \qquad [\mathcal{B}_0, \mathcal{L}] = \frac{1}{6} u_x;$$

setting this into (23) gives

$$u_t + u_x = 0 \tag{23$_0$}$$

$\underline{m = 1}$ $\mathcal{B}_1 = \partial^3 + 2b\partial + b'$; by a short calculation

$$[\mathcal{B}, \mathcal{L}] = \partial\left(\frac{1}{2} u_x - 4 b_x\right)\partial + \frac{1}{6} u_{xxx} - b_{xxx} + \frac{1}{3} b u_x.$$

Choose $b = u/8$; then

$$[\mathcal{B}_1, \mathcal{L}] = (u_{xxx} + u u_x)/24. \tag{23$_1$}$$

Setting into (23) we get, after rescaling, the KdV equation.

Gardner has shown that for every $m > 1$ there is an antisymmetric operator $\mathcal{B}_m$ of order $2m + 1$ such that $[\mathcal{B}_m, \mathcal{L}]$ is of order zero. The zero order term is ∂G_m, G_m the gradient of the Hamiltonian $H^{(m)}$ described in Theorem 1.

It follows from the above result and Theorem 3 that the eigenvalues of $\mathcal{L}(u)$ are conserved under all flows of the KdV hierarchy. We present now another demonstration of this property.

Theorem 5:

i) Let $\lambda = \lambda(u)$ be any of the eigenvalues of $\mathcal{L}(u)$ defined by (22); let $H^{(k)}$ be any of the Hamiltonians in the hierarchy. Then λ and $H^{(k)}$ are in involution:

$$\{\lambda, H^{(k)}\} = 0. \tag{24}$$

ii) Let λ and μ be any two eigenvalues of $\mathcal{L}$. Then λ and μ are in involution

$$\{\lambda, \mu\} = 0 \tag{24$'$}$$

Proof: We compute the gradient of λ with respect to u. Setting $u = u(\epsilon)$ and denoting differentiation with respect to ϵ by $\cdot$ we get from

$$\mathcal{L} w = \lambda w \tag{25}$$

that

$$\mathcal{L}\dot{w} + \dot{\mathcal{L}} v = \lambda \dot{w} + \dot{\lambda} w.$$

Taking the scalar product with w and deducing from (22) that $\dot{\mathcal{L}} = -\dot{u}/6$ we get

$$\dot{\lambda} = -\frac{1}{6}(v, w^2).$$

This shows that

$$G_\lambda = -\frac{1}{6}w^2, \qquad (25)'$$

where w is the normalized *eigenfunction* for the eigenvalue λ.

Next we need the following folklemma:

Lemma 6: *Let w be an eigenfunction λ an eigenvalue of $\mathcal{L}$ defined by (22), and let $\mathcal{N}$ denote the third order operator defined by (9); w^2 satisfies*

$$\mathcal{N}w^2 = 4\lambda\partial w^2. \qquad (26)$$

Proof: by an explicit calculation. $\qquad\qquad\qquad\qquad\qquad\qquad\qquad\square$

We use now definition (3) of the bracket, formula (25)' for G_λ, relation $(13)_k$ and (26):

$$\begin{aligned}
6\{\lambda, H^{(k)}\} &= (6G_\lambda, \partial G_k) = (w^2, \partial G_k) \\
&= (w^2, \mathcal{N}G_{k-1}) = -(\mathcal{N}w^2, G_{k-1}) = \\
&= -4\lambda(\partial w^2, G_{k-1}) = 4\lambda(w^2, \partial G_{k-1}) = \ldots = \\
&= (4\lambda)^{k+1}(w^2, \partial G_{-1}) = 0,
\end{aligned}$$

since by (16), $G_{-1} = -3$.

ii) To prove (24)', denote by z the eigenfunction of μ. Using (26) twice, and the antisymmetry of $\mathcal{N}$ and ∂ we get

$$\begin{aligned}
36\{\lambda, \mu\} &= (6G_\lambda, \partial 6G_\mu) = (w^2, \partial z^2) \\
&= \frac{1}{4\lambda}(w^2, \mathcal{N}z^2) = -\frac{1}{4\lambda}(\mathcal{N}w^2, z^2) \\
&= -\frac{\mu}{\lambda}(\partial w^2, z^2) = \frac{\mu}{\lambda}(w^2, \partial z^2).
\end{aligned}$$

If $\mu \neq \lambda$, this shows that $(w^2, \partial z^2) = 0$; even more so when $w = z$. $\qquad\square$

The conserved quantities $H^{(k)}$ and λ_n are not independent functionals. In the periodic case, when $\mathcal{L}$ has a complete set of eigenfunctions, McKean and van Moerbeke have shown that the $H^{(k)}$ are determined by the eigenvalues, in terms of the asymptotic expansion of λ_n:

$$\lambda_n = n^2 + K^{(0)} + \frac{1}{n^2}K^{(1)} + \ldots \qquad (27)$$

in case of period π.

In the case of the whole line the bulk of the spectrum of $\mathcal{L}$ is continuous when $u(x)$ tends to zero as $x \to \pm\infty$, although there may be a finite number of negative point eigenvalues. These play a most interesting role for the *solitary waves* or *solitons*

of the KdV equation. These are travelling wave solutions of KdV that tend to zero as $x \to \pm\infty$:
$$u(x, t) = s(x - ct).$$
Setting this into (8) yields an ODE for s, explicitely solvable:

$$-cs' + ss' + s''' = 0. \tag{28}$$

Integrate this; since s and all its derivative vanish at ∞,

$$-cs + \frac{1}{2}s^2 + s'' = 0. \tag{28'}$$

Multiply by $2s'$ and integrate once more

$$-cs^2 + \frac{1}{3}s^3 + s'^2 = 0. \tag{28''}$$

The equation can be solved by quadrature, expressible in terms of elementary functions:
$$s(x) = 3c \operatorname{sech}^2 \frac{1}{2} x \sqrt{c}.$$
So there is a one-parameter family of solitary waves, one for each *positive* speed c. The wave is symmetric about its peak, and its peak value is $3c$.

Lemma 7: *Let $s(x)$ be a solitary wave, $\mathcal{L}(s)$ the operator*

$$-\partial^2 - \frac{1}{6}s.$$

$\mathcal{L}(s)$ *has exactly one negative eigenvalue* $\lambda = -\frac{1}{4}c$, *t with eigenfunction* $w = \sqrt{s}$.

Proof: It is a simple calculation using (28)$'$ and (28)$''$ to verify that $w = \sqrt{s}$ is an eigenfunction of $\mathcal{L}(s)$ with eigenvalue $-c/4$. To show why w is so related to s, and that there are no other eigenvalues we resort to the following curious argument pertaining to a very general class of nonlinear evolution equations:

$$u_t = \mathcal{A}(u), \tag{29}$$

u a function on $\mathbb{R}$, also depending on t, $\mathcal{A}$ some nonlinear operator that commutes with translation. $\mathcal{A}$ is assumed differentiable:

$$\frac{d}{d\epsilon} \mathcal{A}(u + \epsilon v) \bigg|_{\epsilon=0} = \mathcal{V}(u)v,$$

where $\mathcal{V}$ is a linear operator. The linearized evolution equation is

$$v_t = \mathcal{V}(u)v, \tag{29'}$$

We assume that the initial value problem for (29) can be solved, uniquely, for a set of initial functions that are dense in L^2, and that (29) is *energy conserving*, i.e. $H^{(0)} = (u, u)$ is a conserved quantity:

$$\frac{d}{dt}(u, u) = 2(u_t, u) = 2(\mathcal{A}(u), u) = 0.$$

Taking u to depend on a parameter ϵ and differentiating this relation with respect to ϵ gives, with $du/d\epsilon = v$,

$$
\begin{aligned}
0 &= (\mathcal{A}(u), v) + (\mathcal{V}(u)v, u) = \\
&= (\mathcal{A}(u) + \mathcal{V}^*(u)u, v).
\end{aligned}
$$

Since the initial values of v are dense,

$$\mathcal{A}(u) + \mathcal{V}^*(u)u = 0. \tag{30}$$

Since $\mathcal{A}(u)$ commutes with translation it is reasonable to look for traveling solitary wave solution $u = s(x - ct)$; setting this into (29) gives for $t = 0$,

$$-cs_x = \mathcal{A}(s).$$

Expressing $\mathcal{A}$ from (30) we can rewrite this as

$$(\mathcal{V}^*(s) - c\partial)s = 0. \tag{31}$$

Let H be any conserved quantity for the evolution equation (29), other than $H^{(0)}$, $G_H = G$ its gradient. We recall that if Q is a *quadratic* conserved quantity for a *linear* evolution equation, then $Q(u, v)$ is conserved for any pair of solutions. The analogue of this for nonlinear evolution equations is this:

If u is any solution of (29) and v any solution of (29)′, then

$$(G(u(t)), v(t)) \tag{32}$$

is independent of t. This follows by differentiating $H(u(t, \epsilon))$ with respect to ϵ, where $u(\epsilon)$ is a one-parameter family of solutions of (29).

Now take $u(t) = s(x - ct)$ in (32) and change variables; denoting $v(x + ct, t) = z(x, t)$ we get that

$$(G(s(x - ct)), v(x, t)) = (G(s), z(t)) \tag{32′}$$

is independent of t. Differentiate (32)′ with respect to t; using $z_t = cv_x + v_t$ and equation (29)′ we get, at $t = 0$,

$$(G(s), (c\partial + \mathcal{V}(s))v) = 0.$$

Transposition gives

$$((\mathcal{V}^*(s) - c\partial)G(s), v) = 0.$$

Since the initial values of v are dense,

$$(\mathcal{V}^*(s) - c\partial)G(s) = 0;$$

in words: $G(s)$ belongs to the nullspace of the operator $\mathcal{V}^*(s) - c\partial$. According to (31) so does s. Let's take it—easily verified for the KdV equation—that the nullspace of $\mathcal{V}^*(s) - c\partial$ in L^2 is one dimensional; then $G(s)$, assumed to be L^2, must be a constant multiple of s:

$$G(s) = \text{const. } s. \tag{33}$$

Take as the conserved quantity $H(u) = \lambda(u)$, an eigenvalue of $\mathcal{L}(s)$. According to (25)', $G_\lambda(u) = -\frac{1}{6}w^2$, where w is the eigenfunction. It follows from (33) that $w = \text{const.} \sqrt{s}$; this shows that all eigenfunctions of $\mathcal{L}(s)$ are equal. Therefore there is only one. $\qquad\qquad\qquad\qquad\qquad\qquad\qquad\qquad\qquad\qquad\qquad\qquad\qquad\qquad\quad$ □

Kruskal and Zabusky have observed in numerical calculations that a number of solitary waves *emerge* from any solution of KdV. That is, *for any ϵ and any L there is a T such that for $t > T$*

$$u(x,t) \qquad and \qquad s(x - c_j t - \varphi_j) \qquad\qquad (34)$$

differ in $-L < x - c_j t < L$ by less than ϵ in the max *and L^2 norm. The same is true for $t < -T$, except that the phase shifts φ_j are different.*

The number of the solitary waves, called in this context solitons, equals the number of negative eigenvalues of $\mathcal{L}(u)$, and the eigenspeeds c_j and eigenvalues λ_j are related by

$$c_j = -4\lambda_j. \qquad\qquad (35)$$

Suppose the asymptotic description (34) in terms of solitons holds; we show that then $-\frac{1}{4}c$ is an eigenvalue of $\mathcal{L}(u)$. For then $u(x+ct,t)$ differs by $< \epsilon$ from $s(x-\varphi)$, so that by Lemma 7 $w(x) = s^{1/2}(x - ct - \varphi)$ is an approximate eigenfunction of $\mathcal{L}(u(t))$ in the sense that

$$\|\mathcal{L}(u(t))w + \frac{1}{4}cw\| \leq \epsilon\|w\|.$$

It follows by spectral theory that $-\frac{1}{4}c$ differs by $< \epsilon$ from an eigenvalue of $\mathcal{L}(u(t))$. Since these eigenvalues are independent of t, and $\epsilon \to 0$ as $t \to \infty$, we conclude that $-\frac{1}{4}c$ is an eigenvalue of $\mathcal{L}(u)$.

To deduce the converse, that an eigenvalue λ indicates the emergence of a soliton with speed $c = -4\lambda$ will be shown in Chapter III, following Gardner, Green, Kruskal, Miura.

Chapter II. Scattering and inverse scattering

Denote, as before, by $\mathcal{L}$ a second order operator

$$\mathcal{L} = -\partial_x^2 + q$$

acting on the whole line $\mathbb{R}$, $q(x)$ a potential that is smooth and dies down fast to 0 as $x \to \pm\infty$.

The interval $[0, \infty)$ belongs to the *continuous spectrum* of $\mathcal{L}$, of multiplicity two. The generalized eigenfunctions are solutions of

$$\mathcal{L}f = k^2 f, \qquad k \text{ real.} \tag{1}$$

Since we have assumed that $q(x)$ tends to 0 rapidly as $|x| \to \infty$, it follows easily that for $|x|$ large, solutions f of (1) are very nearly superpositions of e^{ikx} and e^{-ikx}. It is convenient to choose as basis of solutions of (1) the *Jost functions*, characterized by

$$
\begin{aligned}
f_+(x, k) &\simeq e^{ikx} &&\text{as } x \to +\infty, \\
f_-(x, k) &\simeq e^{-ikx} &&\text{as } x \to -\infty.
\end{aligned}
\tag{2}
$$

These are complex valued solutions; clearly

$$\overline{f_+}(k) = f_+(-k), \qquad \overline{f_-}(k) = f_-(-k). \tag{3}$$

Since any three solutions of (1) are linearly dependent, there must be for $k \neq 0$ relations of the form

$$f_-(k) = a_+(k)\overline{f_+}(k) + b_+(k)f_+(k) \tag{4$_+$}$$

and

$$f_+(k) = a_-(k)\overline{f_-}(k) + b_-(k)f_-(k). \tag{4$_-$}$$

Properties of the functions $a_+, a_-, b_+.b_-$ and relations among them can be deduced using the Wronskian bilinear form, defined as

$$W(f, g) = \det \begin{pmatrix} f & g \\ f' & g' \end{pmatrix} = fg' - f'g. \tag{5}$$

Clearly, W is *antisymmetric*. For a pair of solutions of equation (1), W is independent of x; in particular for the Jost functions

$$
\begin{aligned}
W(f_+(k), \overline{f_+}(k)) &= -2ik, \\
W(f_-(k), \overline{f_-}(k)) &= 2ik.
\end{aligned}
\tag{5$'$}
$$

can be evaluated at x near $+\infty$, resp. $-\infty$.

Form the Wronskian of (4)$_+$ with $f_+(k)$, resp. $\overline{f_+}(f)$; using (5)$'$ we get

$$
\begin{aligned}
W(f_-(k), f_+(k))/2ik &= a_+(k), \\
W(f_-(k), \overline{f_+}(k))/2ik &= -b_+(k).
\end{aligned}
\tag{6$_+$}
$$

Expressing f_- and f_+ in (5)$'$ by (4)$_+$ gives

$$1 = |a_+|^2 - |b_+|^2. \tag{6$'_+$}$$

Similarly we get

$$W(f_-(k), f_+(k))/2ik = a_-(k),$$
$$W(\overline{f_-}(k), f_+(k))/2ik = -b_-(k) \tag{6}_-$$

and

$$1 = |a_-|^2 - |b_-|^2. \tag{6}'_-$$

An important conclusion is that

$$a_-(k) = a_+(k),$$

henceforth denoted as $a(k)$.

It is natural to divide relations $(4)_\pm$ by $a(k)$ and write them as

$$T(k)f_-(k) = \overline{f_+}(k) + R_+(k)f_+(k), \tag{7}_+$$

and

$$T(k)f_+(k) = \overline{f_-}(k) + R_-(k)f_-(k); \tag{7}_-$$

here T and $R_\pm$ abbreviate

$$T(k) = \frac{1}{a(k)}, \qquad R_\pm(k) = \frac{b_\pm(k)}{a(k)}. \tag{8}$$

In view of the asymptotic behaviour (2) of f_+ and f_- near $x = \pm\infty$ we can think of $(7)_+$ as a harmonic wave coming from $+\infty$, being partly transmitted, partly reflected by the potential q. The function $T(k)$ is called the *transmission coefficient*, $R_+(k)$ and $R_-(k)$ *reflection coefficients* to the right and to the left respectively.

Divide $(6)'_\pm$ by $|a(k)|^2$; we get

$$|T(k)|^2 + |R_\pm(k)|^2 = 1, \tag{9}$$

a form of *conservation of energy* for transmitted and reflected waves.

Formulas (3) and (6) show that $R_+(k)$ and $R_-(k)$ defined by (8) are *skew symmetric* functions:

$$R_+(-k) = \overline{R_+(k)}, \qquad R_-(-k) = \overline{R_-(k)}. \tag{10}$$

For k large the term qf in equation (1) is negligible compared to $k^2 f$, therefore most of the signed coming from ∞ will be transmitted, very little reflected. This can be expressed as follows

$$\lim_{k\to\infty} R_\pm(k) = 0, \qquad \lim_{k\to\infty} T(k) = 1. \tag{11}$$

We turn now to the asymptotic behaviour of $f_+(x, k)$ for k large. We set

$$f_+(x, k) = e^{ikx + p(x,k)}. \tag{12}$$

According to equation (1), p satisfies

$$p'' + 2ikp' + p'^2 = q.$$

We take p to be

$$p \simeq \frac{p_1}{k} + \frac{p_2}{k^2} + \cdots \tag{13}$$

Then

$$p_1' = \frac{q}{2i}, \qquad p_2 = \frac{q}{4}, \dots \qquad (14)$$

We turn now to *complex* values of k in the upper half plane (UHP), $\operatorname{Im} k > 0$. It is not hard to show that the Jost solutions are well defined for $\operatorname{Im} k > 0$, and that $f_\pm(x, k)$ are *analytic functions* of k. The growth of the Jost functions is easily estimated: $|f_+(x, k)|$ is roughly like $e^{-\operatorname{Im} kx}$, $|f_-(x, k)|$ like $e^{\operatorname{Im} kx}$. The following more precise estimate, not hard, plays an important role:

For each x, $e^{-ikx} f_+(x, k) - 1$ and $e^{ikx} f_-(x, k) - 1$ as functions of k belong to the Hardy class H_+^2.

H_+^2 consists of functions analytic in the UHP, whose square integrals along all lines $\operatorname{Im} k = \text{const} > 0$ are bounded by some constant.

Formula $(6)_+$ shows that $a(k)$ is an analytic function in the UHP. It is easy to show that $a(k) \to 1$ as $k \to \infty$ in the UHP. The zeros of $a(k)$ in the UHP play an important role. Suppose $a(n) = 0$; by formula (6) for $a(n)$,

$$W(f_+(n), f_-(n)) = 0.$$

The vanishing of Wronskian implies linear dependence

$$d f_+(n) = f_-(n). \qquad (15)$$

When $\operatorname{Im} n > 0$, (2) shows that $f_+(x, n)$ decays exponentially as $x \to +\infty$, and $f_-(x, n)$ decays as $x \to -\infty$. Therefore by (15), $f_+(n)$ is square integrable on $\mathbb{R}$, and so a proper eigenfunction of $\mathcal{L}$, with eigenvalue n^2. Since $\mathcal{L}$ is self-adjoint, its eigenvalues are real; this shows that the zeros n of $a(k)$ lies on the imaginary axis: $n = i\eta$.

Next we calculate the derivative of $a(k)$ at n. Differentiate (6) with respect to k; writing $\dfrac{d}{dk} = \dot{}$ we get at $k = i\eta$

$$\dot{a}(i\eta) = \frac{-1}{2\eta} \{W(\dot{f}_-, f_+) + W(f_-, \dot{f}_+)\}.$$

We shall evaluate the right side at any point x whatever. Setting (15) into this formula gives

$$\dot{a}(i\eta) = \frac{-1}{2\eta} \{d^{-1} W(\dot{f}_-, f_-) + d W(f_+, \dot{f}_+)\}. \qquad (16)$$

The functions $f_\pm$ satisfy equation (1):

$$-f_{xx} + qf = k^2 f. \qquad (17)$$

Differentiate with respect to k:

$$-\dot{f}_{xx} + q\dot{f} = k^2 \dot{f} + 2kf. \qquad (17)'$$

Multiply (17) by $\dot{f}$ and (17)' by f, and subtract the two; we get

$$(\dot{f} f_x - f \dot{f}_x)_x = 2k f^2. \qquad (17)''$$

The bracket on the left is $W'(\dot{f}, f)$. Now we set $f = f_+$ and f_-; recalling that for $\operatorname{Im} k > 0$, f_+ vanishes at $+\infty$, f_- at $-\infty$, we integrate $(17)''_{\pm}$ from $\pm\infty$ to x; we get

$$W(\dot{f}_+, f_+)(x) = -2k \int_x^\infty f_+^2\, dx,$$

$$W(\dot{f}_-, f_-)(x) = 2k \int_{-\infty}^x f_-^2\, dx.$$

Now set $k = i\eta$ and substitute the above relations into (16). Using (15) we can eliminate f_- to obtain the formula

$$\dot{a}(i\eta) = -id \int_{-\infty}^\infty f_+^2(x, i\eta)\, dx. \tag{18}$$

Denote the Fourier transform of $f_+(x, k) - e^{ikx}$ with respect to k by $B_+(x, y)$. We shall omit the subscript $+$ when we can do so without causing confusion. Since by (3), $f_+(k) - e^{ikx}$ is a skew symmetric function of k, B is real. We have noted before that f_+ is an analytic function of k in the UHP, and that $f_+(x, k)e^{-ikx} - 1$ belongs to Hardy class H_+^2. According to the Paley-Wiener theorem the Fourier transform of such a function is supported on the positive axis. It follows then that the Fourier transform of $f_+(x, k) - e^{ikx}$ is supported on $[x, \infty)$:

$$f_+(x, k) = e^{ikx} + \int_x^\infty B(x, y)e^{iky}\, dy. \tag{19}$$

The PW theorem guarantees that B is a square integrable function of y; under our generous assumptions about the potential q it is even C^1. Integrating by parts on the right we get

$$f_+(x, k) = e^{ikx} - B(x, x)\frac{e^{ikx}}{ik} + \frac{i}{k}\int B_y(x, y)e^{iky}\, dy. \tag{19'}$$

On the other hand, it follows from the asymptotics (12) and (13) that

$$\begin{aligned} f_+(x, k) &= e^{ikx}(1 + p + \ldots) = \\ &= e^{ikx}\left(1 + \frac{p_1}{k} + \ldots\right) = e^{ikx} + \frac{e^{ikx}p_1}{k} + \ldots \end{aligned}$$

Comparing this with (19)′ we conclude that $iB(x, x) = p_1(x)$. Differentiating and using (14) we get

$$q(x) = -2\frac{d}{dx}B(x, x). \tag{20}$$

We return now to relation $(7)_+$:

$$\overline{f_+} + R_+(k)f_+ = T(k)f_-. \tag{7}_+$$

Uniqueness Theorem 1: *Suppose q_1 and q_2 are two potentials such that the associated Schroedinger operators $\mathcal{L}_1$ and $\mathcal{L}_2$ have no point eigenvalues, and suppose that* *the right reflection coefficient $R_+(k)$ is the same for both operators. Then $\mathcal{L}_1 \equiv \mathcal{L}_2$. i.e. $q_1 \equiv q_2$.*

In short: For Schroedinger operators without point spectrum the potential is uniquely determined by the right reflection coefficient.

Proof (following Deift and Trubowitz): Abbreviate

$$e^{-ikx}f_+ - 1 = e_+, \qquad e^{ikx}f_- - 1 = e_-. \tag{21}$$

Multiply $(7)_+$ by e^{ikx} and use (21); for k real we get

$$\overline{e_+} + 1 + R_+(k)e^{2ikx}(e_+ + 1) = T(k)(e_- + 1).$$

Such an identity holds for both $\mathcal{L}_1$ and $\mathcal{L}_2$; subtract them. Using the obvious notation

$$\Delta e = e_1 - e_2, \quad \text{etc.}$$

we write

$$\Delta\overline{e_+} + R_+ e^{2ikx}\Delta e_+ = \Delta T + \Delta(Te_-). \tag{22}$$

We saw earlier that e_+—and therefore Δe_+—is a Hardy class H_+^2 function. By assumption of no point spectrum the analytic function $a(k)$ has no zeros in the UHP, so by (8) and (11) $T = 1/a$ is analytic and bounded there. The estimate (11) can be refined to show that in fact $T(k) = 1 + O(1/k)$ for $|k|$ large. This shows that ΔT belongs to H_+^2. Since T is bounded and e_- belongs to H_+^2, so does Te_-. This shows that the right side of (22) belongs to H_+^2.

Multiply (22) by Δe and integrate with respect to k over $\mathbb{R}$. On the right we use Cauchy's theorem to shift the path of integration to $\text{Im } k = K$ and let $K \to \infty$. It is a well-known property of Hardy class functions that their square integral along $\text{Im } k = K$ tends to zero as $K \to \infty$. So we obtain from (22) that

$$\int\limits_{-\infty}^{\infty} |\Delta e_+|^2 \, dk + \int\limits_{-\infty}^{\infty} R_+(k)e^{2ikx}(\Delta e_+)^2 \, dk = 0. \tag{22$'$}$$

It follows from (9) that $|R_+(k)| \leq 1$, and from (11) that the strict inequality holds for k large. This shows that the second integral in (22)' is strictly less in absolute value than the first; this can be only if the square integral of $|\Delta e_+|$, and so Δe_+ itself, is zero. But then by (21), $\Delta f_+ = 0$, which implies that $\Delta q = 0$. □.

Next we go beyond mere uniqueness and show how to reconstruct q from R_+. Set the Fourier representations of f_+, f_- and R_+ into $(7)_+$. For f_+ we have (19), for f_- analogously

$$f_-(x, k) = e^{-ikx} + \int\limits_{-x}^{\infty} B_-(x, y)e^{iky} \, dy$$

while

$$R_+(k) = \int\limits_{-\infty}^{\infty} r(y)e^{iky} \, dy.$$

Since by (10), R_+ is skew symmetric, its F.T. r is real.

We get in $(7)_+$

$$e^{-ikx} + \int\limits_x^\infty B(x,y)e^{-iky}\,dy + \int\limits_{-\infty}^\infty r(y)e^{ik(y+x)}\,dy +$$

$$\int r(z)e^{ikz}\,dz \int\limits_x^\infty B(x,y)e^{iky}\,dy = \tag{6}$$

$$= T(k)e^{-ikx} + T(k)e_-(x,k)e^{-ikx}.$$

We rewrite this equations as

$$\int\limits_{-\infty}^{-x} B(x,-y)e^{iky}\,dy + \int\limits_{-\infty}^\infty r(y-x)e^{iky}\,dy +$$

$$\int r * B(x)e^{iky}\,dy = \tag{6}$$

$$(T(k)-1)e^{-ikx} + T(k)e_-(x,k)e^{-ikx}.$$

We claim that the Fourier transform of the right side is supported on $[-x,\infty)$. For we saw earlier that $T(k)-1$ belongs to Hardy class H^2_+ and so its F.T. lies on $[0,\infty)$, and that $T(k)e_-$ is of Hardy class. Multiplication by e^{-ikx} shifts the support of their F.T. to $[-x,\infty)$. So we conclude that the F.T. of the left side of $(23)'$ is zero for $y \le -x$:

$$B(x,-y) + r(y-x) + \int\limits_x^\infty r(y-z)B(x,z)\,dz = 0. \tag{23''}$$

for $y < -x$. We replace finally y by $-y$, and denote $r(-z)$ as $P(z)$. Then $(23)''$ can be written as

$$B(x,y) + \int\limits_x^\infty P(y+z)B(x,z)\,dz + P(y+x) = 0 \tag{24}$$

for $x \le y$.

Equation (24) is an *inhomogeneous Fredholm integral equation* for B; the variable x is merely a *parameter*.

(24) is the celebrated *Gelfand-Levitan-Marchenko* equation. Combined with formula (20) its solution B furnishes the value of the potential q in terms of the reflection coefficient R_+.

What is remarkable about this process of reconstruction of q is that it is in terms of *linear processes*.

We turn now to the case when the Schroedinger operator has points spectrum (called *bound states* in physics). In this case the function $a(k)$ has zeros in the UHP, so $T(k) = a(k)^{-1}$ has poles there. These poles are located on the imaginary axis at $i\eta$, $\eta = \eta_1, \ldots, \eta_N$. In this case the Fourier transform of the right side of $(23)'$ has nonzero contribution when $y < -x$. Its value can be found by shifting the path of integration in the Fourier integral

$$\frac{1}{2\pi} \int\limits_{-\infty}^\infty T(k)f_-(k)e^{-iky}\,dk \tag{25}$$

from k real to $\operatorname{Im} k = K$, $K \to \infty$. The residue of the integrand at $k = i\eta$ is

$$\frac{1}{\dot{a}(i\eta)} f_-(i\eta) e^{\eta y}. \tag{25}'$$

The value of $\dot{a}$ is furnished by (18), while according to (15), $f_-(i\eta) = df_+(i\eta)$. Setting these in $(25)'$ and multiplying by $2\pi i$ we get the following value for (25) when $y < -x$:

$$-\sum \frac{f_+(i\eta) e^{\eta y}}{\int f_+^2(i\eta)\, dx}. \tag{25}''$$

We substitute the Fourier representation (19) of f_+ into $(25)''$. Using the abbreviation

$$C(\eta) = \Big(\int\limits_{-\infty}^{\infty} f_+^2(i\eta) \Big)^{-1/2} \tag{26}$$

we can using (19) rewrite $(25)''$ as

$$-\sum C^2(\eta) \Big[e^{-\eta(x-y)} + \int\limits_{x}^{\infty} B(x,z) e^{-\eta(z-y)}\, dz \Big].$$

Adding this sum to the Fourier transform of $(23)'$ and changing y to $-y$ we obtain an equation of the *same form as (24)*, except that P is now redefined as

$$P(z) = r(-z) + \sum C^2(\eta) e^{-\eta z}. \tag{27}$$

This is the complete form of the G-L-M equation.

The η, $C(\eta)$ appearing in equation (27) have this meaning: $-\eta^2$ is an eigenvalue of $\mathcal{L}$, and $C(\eta)$ is the $(+)$ *norming constant*, so called because $C(\eta) f_+(i\eta)$ has L^2 norm $= 1$, as is evident from definition (26).

The complete form of the uniqueness theorem says that *the potential q is uniquely determined by the right reflection coefficient $R_+(k)$, the eigenvalues $-\eta_j^2$ and the right norming constants C_j.*

This is equivalent to saying that the Fredholm operator (24), with P defined by (27), is invertible for every x. We omit the proof.

We turn now to another extreme case, *reflectionless potentials*, i.e. for which the right reflection coefficient $R_+(k)$ is zero for all k. It follows incidentally from (19) that for such potentials also $R_- \equiv 0$. For such potentials, discussed by Kay and Moses in the '50-ies, the function P given by (27) contains only the exponential terms; therefore the kernel $P(y+z)$ is degenerate and so the G-L-M equation (24) can be reduced to a finite system of linear equation, as follows:

When $P(z)$ is a finite linear combination of exponentials,

$$P(z) = \sum C_j^2 e^{-\eta_j z}, \tag{27}'$$

the second and third term in (24) are for fixed x also linear combinations of exponentials in y; therefore so is $B(x,y)$ itself. To maintain symmetry, we write

$$B(x,y) = \sum b_i(x) C_i e^{-\eta_i y}, \tag{28}$$

b_i functions to be determined from (24). Setting (27)' and (28) into (24) and carrying out the z integration we obtain

$$\sum b_i C_i e^{-\eta_i y} + \sum \frac{C_j^2 b_i C_i e^{-\eta_j y - (\eta_i + \eta_j)x}}{\eta_i + \eta_j}$$
$$+ \sum C_i^2 e^{-\eta_i(x+y)} = 0.$$

Equating to zero the exponential terms $e^{-\eta_a y}$ occurring gives, after division by C_i,

$$b_i + \sum \frac{C_i C_j e^{-(\eta_i + \eta_j)x}}{\eta_i + \eta_j} b_j + C_i e^{-\eta_i x} = 0. \tag{29}$$

We solve this system of equations by Cramer's rule:

$$b_i = -\sum_j \frac{D_{ij}}{D} C_j e^{-\eta_j x},$$

where D is the determinant of (29):

$$D = \det\left(I + \frac{C_i C_j}{\eta_i + \eta_j} e^{-\eta_i x} e^{-\eta_j x}\right) \tag{29}'$$

and D_{ij} are the signed minors of D of co-order 1. Setting this into (28) we get

$$B(x, y) = \frac{-1}{D} \sum D_{ij} C_i C_j e^{-\eta_i y} e^{-\eta_j x}.$$

For $y = x$ this can be rewritten in the elegant form

$$B(x, x) = \frac{d}{dx} \log D(x). \tag{30}$$

This can be verified immediately by differentiating D as given by (29)', using the rule for differentiating a determinant by, say, columns, and expanding each of the determinants in the resulting sum of determinants by its columns.

Dyson has observed that formula (30) holds for the solution of the G-L-M equation (24) even when P does not come from a reflectionless potential. In this case D has to be interpreted as the *Fredholm determinant* of the Fredholm operator on the left in (24). Since x enters (24) as a parameter, D depends on x. Setting (30) into formula (20) for q we get the following remarkable formula for the potential:

$$\boxed{q(x) = -2\frac{d^2}{dx^2} \log D(x))} \tag{30}'$$

We turn now, briefly, to the case when the potential q, instead of tending to 0 as $x \to \pm\infty$, is *periodic* in x. In this case the operator $\mathcal{L}$, acting on periodic functions, can be regarded as a selfadjoint operator on $L^2[0, 2\pi]$. Unlike for the whole real line, the spectrum of $\mathcal{L}$ is pure point spectrum. The reconstruction of the potential in this situation is based on the pointspectrum of $\mathcal{L}$ over the periodic functions *and* its pointspectrum under Dirichlet boundary conditions. More precisely:

Denote by $\lambda_0^P, \lambda_1^P, \ldots$ the eigenvalues of $\mathcal{L}$ under *periodic* boundary conditions:

$$w(0) = w(2\pi), \qquad w'(0) = w'(2\pi). \tag{31}_P$$

Denote by $\lambda_1^A, \ldots$ the eigenvalues of $\mathcal{L}$ under *antiperiodic* boundary conditions

$$-w(0) = w(2\pi), \qquad -w'(0) = w'(2\pi). \tag{31}_A$$

Denote by $\lambda_1^D(y), \ldots$ the eigenvalues of $\mathcal{L}$ under *Dirichlet* boundary conditions:

$$w(y) = 0, \qquad w(y + 2\pi) = 0. \tag{31}_D$$

Theorem (trace formula):

$$\boxed{q(y) = \lambda_0^P + \sum(\lambda_j^P + \lambda_j^A - 2\lambda_j^D(y))} \tag{32}$$

This remarkable formula is due to Gelfand and Levitan; the following simple, intuitive derivation has been given by Lax:

Assume, at the cost of subtracting a constant from q, that

$$\int_0^{2\pi} q(x)\, dx = 0. \tag{33}$$

Deform the potential through a smooth one-parameter family of potentials $q(x, t)$ to zero. That is

$$q(x, 1) = q(x), \qquad q(x, 0) \equiv 0.$$

Take case that (33) is satisfied during deformation, differentiating with respect to t, denoted as $\dot{}$, we get

$$\int \dot{q}\, dx = 0. \tag{33}'$$

We calculate now the rate of change of an eigenvalue during deformation. Differentiating

$$\mathcal{L}w = \lambda w$$

with respect to t gives

$$\mathcal{L}\dot{w} + \dot{\mathcal{L}}w = \lambda\dot{w} + \dot{\lambda}w.$$

Taking the scalar product with w and using the selfadjointness of $\mathcal{L}$ we get

$$(\dot{\mathcal{L}}w, w) = \dot{\lambda},$$

assuming w to have norm 1. Since only q changes during deformation, we can rewrite this as

$$\int_0^{2\pi} \dot{q}w^2\, dx = \dot{\lambda}.$$

Since condition (33)' holds, we may rewrite this as

$$\int_0^{2\pi} \dot{q}\left(w^2 - \frac{1}{2\pi}\right) = \dot{\lambda}. \tag{34}$$

Integrating this with respect to t gives

$$\int\limits_0^1 \int\limits_0^{2\pi} \dot{q}\left(w^2 - \frac{1}{2\pi}\right) dx\,dt = \lambda - \lambda(0),$$

where $\lambda(0)$ is the corresponding eigenvalue of $\mathcal{L}(0) = -\partial_x^2$.

This relation holds for each eigenvalue. We sum over all eigenvalues, assuming that the sums converge, at least in the sense of distributions:

$$\iint \dot{q} \sum \left(w_n^2 - \frac{1}{2\pi}\right) dx\,dt = \sum \lambda_n - \lambda_n(0). \tag{35}$$

Next we claim that the sum

$$S = \sum w_n^2 - \frac{1}{2\pi} \tag{36}$$

is independent of the deformation parameter t. For differentiate (36) termwise:

$$\dot{S} = \sum \frac{d}{dt}(w_n^2 - 2\pi) = 2\sum w_n \dot{w}_n.$$

Expand $\dot{w}_n$ as

$$\dot{w}_n = \sum a_{nm} w_m, \qquad a_{nm} = (\dot{w}_n, w_m).$$

Then

$$\dot{S} = 2\sum a_{nm} w_n w_m. \tag{37}$$

Since the $\{w_n\}$ form an orthonornal basis

$$(w_n, w_m) = \delta_{nm}$$

differentiating with respect to t gives

$$(\dot{w}_n, w_m) + (w_n, \dot{w}_m) = 0,$$

i.e. $a_{nm} + a_{mn} = 0$. But then $\dot{S}$ given by (37) is zero.

We can now carry out the t integration in (35); since $q(0) = 0$,

$$\int qS\,dx = \sum \lambda_n - \lambda_n(0). \tag{38}$$

This is true for all boundary conditions under which $\mathcal{L}$ is selfadjoint, in particular the three mentioned above. We evaluate the sum S in each case, using the operator $\mathcal{L}(0) = -\partial^2$.

$$w_0^P = \frac{1}{\sqrt{2\pi}}, \qquad w_{2n-1}^P = \frac{1}{\sqrt{\pi}} \sin nx,$$

$$w_{2n}^P = \frac{1}{\sqrt{\pi}} \cos nx;$$

$$\lambda_0^P(0) = 0, \qquad \lambda_{2n-1} = \lambda_{2n} = n^2. \tag{39}^P$$

Clearly

$$S^P \equiv 0.$$

In the antiperiodic case

$$w^A_{2n-1} = \frac{1}{\sqrt{\pi}} \sin\left(n - \frac{1}{2}\right)x, \quad w^A_{2n} = \frac{1}{\sqrt{\pi}} \cos\left(n - \frac{1}{2}\right)x,$$

$$\lambda_{2n-1}(0) = \lambda_{2n}(0) = (n - 1/2)^2. \tag{39}^A$$

Clearly $S^A \equiv 0$.

In the Dirichlet case

$$w^D_m = \frac{1}{\sqrt{\pi}} \sin\frac{m}{2}x;$$

setting $m = 2n - 1, 2n$ we get

$$\lambda^D_{2n-1}(0) = (n - 1/2)^2, \qquad \lambda^D_{2n} = n^2. \tag{39}^D$$

Using the trig identity $\sin^2 \alpha = \dfrac{1 - \cos 2\alpha}{2}$ we get

$$\begin{aligned} S^D &= \sum_1^\infty \left(\frac{1}{\pi}\sin^2\frac{m}{2}x - \frac{1}{2\pi}\right) = -\frac{1}{2\pi}\sum_1^\infty \cos mx \\ &= -\frac{1}{2}\delta(x) + \frac{1}{4\pi}. \end{aligned}$$

Setting these values into (38) we get

$$0 = \lambda^P_0 + \sum(\lambda^P_{2n-1} + \lambda^P_{2n} - 2n^2) \tag{38}^P$$

$$0 = \sum(\lambda^A_{2n-1} + \lambda^A_{2n} - 2(n - 1/2)^2). \tag{38}^A$$

Since $\int q\,dx = 0$, in the Dirichlet case

$$-\frac{1}{2}q(0) = \sum(\lambda^D_{2n-1} + \lambda^D_{2n} - (n - 1/2)^2 - n^2). \tag{38}^D$$

Adding $(38)^P$ and $(38)^A$ and subtracting $(38)^D$ gives the trace formula (32) at $y = 0$.

$\sqcap$

Chapter III. Solutions of KdV via inverse scattering

We have seen in Chapter I that if u is a solution of the KdV equation (8), then the eigenvalues of the operator $-\partial^2 - u/6$ are independent of t. The norming constants of the eigenfunctions and the reflection coefficient do change with time, but in a simple, explicitly calculable fashion. We derive these formulas now but for sake of conformity with Chapter II we change the coefficient occurring in the KdV equation. We redefine $\mathcal{L}(t)$ to be

$$\mathcal{L}(t) = -\partial^2 + u, \tag{1}$$

$\mathcal{B}(t)$ to be

$$\mathcal{B}(t) = -4\partial^3 + 6u\partial + 6u'. \tag{2}$$

Their commutator is a zero order operator:

$$[\mathcal{B}, \mathcal{L}] = -u''' + 6uu',$$

where $'$ denotes differentiation with respect to x. So the commutator equation

$$\mathcal{L}_t = [\mathcal{B}, \mathcal{L}] \tag{3}$$

is the renormalized KdV equation

$$u_t - 6uu_x + u_{xxx} = 0. \tag{3'}$$

Recall now from Chapter I the one-parameter family of operators $\mathcal{U}(t)$ defined by

$$\mathcal{U}_t = \mathcal{B}\mathcal{U}, \qquad \mathcal{U}(0) = \mathcal{I}. \tag{4}$$

Since $\mathcal{B}$ is antiselfadjoint, $\mathcal{U}(t)$ is a unitary operator for every value of t. It follows from the commutator equation (3) and eqn. (4) that

$$\mathcal{U}^*(t)\mathcal{L}(t)\mathcal{U}(t)$$

is independent of t. Since $\mathcal{U}(0) = \mathcal{I}$, we deduce that

$$\mathcal{L}(t)\mathcal{U}(t) = \mathcal{U}(t)\mathcal{L}(0). \tag{5}$$

Let w_0 be an eigenfunction of $\mathcal{L}(0)$:

$$\mathcal{L}(0)w_0 = -\eta^2 w_0, \qquad \eta > 0. \tag{6}$$

Letting (5) act on w we get

$$\mathcal{L}(t)\mathcal{U}(t)w_0 = -\eta^2\mathcal{U}(t)w_0; \tag{6'}$$

this shows that $\mathcal{U}(t)w_0 = w(t)$ is an eigenfunction of $\mathcal{L}(t)$ with eigenvalue $-\eta^2$. If w_0 has norm 1, so does $w(t)$, since $\mathcal{U}(t)$ is unitary.

Differentiate $w(t) = \mathcal{U}(t)\dot{w}_0$ with respect to t and use (4); we get

$$w_t = \mathcal{U}_t w_0 = \mathcal{B}\mathcal{U}w_0 = \mathcal{B}w. \tag{7}$$

We recall from Chapter II, eqn. (26) that the normalized eigenfunctions of $\mathcal{L}$ have the form

$$w = Cf_+ \tag{8}$$

where

$$f_+(x) \simeq e^{-\eta x} \quad \text{as } x \to +\infty. \tag{8}'$$

Since $u(x) \to 0$ as $x \to \pm\infty$, the asymptotic behaviour of $\mathcal{B}$, given by (2), is

$$\mathcal{B} \simeq -4\partial^3 \quad \text{as } x \to \pm\infty. \tag{8}''$$

Setting these asymptotic derivations into (7) gives

$$\frac{d}{dt}Ce^{-\eta x} \simeq -4\partial^3 Ce^{-\eta x} = 4\eta^3 Ce^{-\eta x},$$

i.e.

$$\frac{d}{dt}C = 4\eta^3 C. \tag{9}$$

Thus

$$C(t) = e^{4\eta^3 t}C(0). \tag{9}'$$

To determine the time evolution of the reflection coefficient we take for k real the eigenfunction

$$f_0 = \overline{f_+} + R_0 f_+ = T f_-.$$

The asymptotic behaviour of f_0 is

$$f_0(x, k) \simeq \begin{cases} e^{-ikx} + R_0(k)e^{ikx}, & x \to +\infty \\ T(k)e^{-ikx}, & x \to -\infty; \end{cases} \tag{10}$$

f_0 satisfies

$$\mathcal{L}(0)f_0 = k^2 f_0.$$

Define $f(t)$ by

$$f(t) = \mathcal{U}(t)f_0. \tag{11}$$

It follows from (5) that $f(t)$ is an eigenfunction of $\mathcal{L}(t)$:

$$\mathcal{L}(t)f(t) = k^2 f(t).$$

The asymptotic behaviour of $f(t)$ as $x \to \pm\infty$ is of the form

$$f(x, k, t) \simeq \begin{cases} de^{-ikx} + ee^{ikx} \\ ge^{-ikx} + he^{ikx}, \end{cases} \tag{10}'$$

where d, e, g and h are functions of k and t. To determine what kind of function, we differentiate (11) with respect to t; using (4) we get

$$f_t = \mathcal{B}f.$$

Replacing f and $\mathcal{B}$ by their asymptotic values (10)' and (8)'' we get four separate equations for d, e, g and h:

$$d_t = -4ik^3 d, \qquad e_t = 4ik^3 e,$$
$$g_t = -4ik^3 g, \qquad h_t = 4ik^3 h.$$

Together with the initial values (10) we get

$$d(t) = e^{-4ik^3t}, \qquad e(t) = R_0(k)e^{4ik^3t},$$
$$g(t) = T(k)e^{-4ik^3t}, \qquad h(t) = 0.$$

Setting this into (10)' we conclude that

$$e^{4ik^3t}f(t) \simeq \begin{cases} e^{-ikx} + R_0(k)e^{8ik^3t}, & x \to +\infty \\ T(k)e^{-ikx}, & x \to -\infty. \end{cases}$$

It follows that

$$R(k,t) = R_0(k)e^{8ik^3t},$$
$$T(k,t) = T_0(k). \tag{12}$$

The solution of the initial value problem for the KdV equation (3)' is now at hand, in three easy steps and a hard one:

Step 1: Calculate the scattering data of the operator $\mathcal{L}_0$ for $u = u_0$, the prescribed initial function. The scattering data consist of the eigenvalues $-\eta_j^2$, $j = 1, \ldots, N$, the corresponding norming are constants $C_j(0)$, $j = 1, \ldots, N$, and the reflection coefficient $R_0(k)$ as function of k.

Step 2: Determine the scattering data of $\mathcal{L}(t)$ as

$$-\eta_j^2, \ C_j(t) = e^{4\eta_j^3 t}C_j(0),$$
$$R(k,t) = R_0(k)e^{8ik^3t}. \tag{13}$$

Step 3: Calculate the Fredholm determinant $D(x,t)$ of the Fredholm operator (24), Chapter II, where $P(z,t)$ is defined by (27), with scattering data given by (13).

Step 4: According to formula (30)' of Chapter II, the potential in $\mathcal{L}(t)$ is given by

$$u(x,t) = -2\frac{d^2}{dx^2}\log D(x,t). \tag{14}$$

The hard step is step 3, for there are no easy ways of calculating Fredholm determinants. However asymptotic evaluation is possible in certain limiting situations, such as for large time, and small dispersion. In both cases, only the contribution of the point spectrum matters.

We return now to the reflectionless case, where the determinant is given by formula (29)' of Chapter II:

$$D = \det\left(I + \frac{C_iC_j}{\eta_i + \eta_j}e^{-\eta_i x - \eta_j x}\right).$$

The time dependence of the norming constants is given by formula (9)'; setting this into the above formula gives D as function of x and t:

$$D(x,t) = \det\left(I + \frac{C_i(0)C_j(0)}{\eta_i + \eta_j}e^{4\eta_i^3 t - \eta_i x}e^{4\eta_j^3 t - \eta_j x}\right). \tag{15}$$

We rewrite the formula, using the following determinant identity: for any matrix M

$$\det(I + M) = \sum \det M_S, \tag{16}$$

the summation being over all principal minors M, indexed by the set S of indices retained in the minor. In our case

$$M_{ij} = \frac{g_i g_j}{\eta_i + \eta_j},$$

where

$$g_i = C_i(0)e^{4\eta_i^3 t - \eta_i x}. \tag{17}$$

We can write

$$M = GHG,$$

where H is the matrix

$$H_{ij} = \frac{1}{\eta_i + \eta_j},$$

and G the diagonal matrix with entries g_i. Similarly, the principal minors of M are of the form

$$M_S = G_S H_S G_S,$$

where the subscripted matrices are obtained from the unsubscripted ones by retaining only those rows and columns whose index belongs to S. Therefore

$$\det M_S = h_S \left(\prod_S g_i \right)^2, \tag{18}$$

where $h_S = \det H_S$.

We shall determine now the asymptotic behaviour of $D(x, t)$ for large positive x and t. We distinguish two cases:

i) $|x - c_j t| > vt$ for all j, where c_j denote the soliton speeds associated with the initial data, and v is any positive constant.

ii) $|x - c_K t| < d$ for some K

In Chapter I we saw that the soliton speeds c for the KdV equation are -4λ, where λ are the eigenvalues of the associated operator $\mathcal{L}$ defined by (22). The same relation holds for the renormalized KdV equation (3)' and the operator (1):

$$u_t - 6uu_x + u_{xxx} = 0. \tag{3'}$$

Setting $u = s(x - ct)$ into (3)' yields

$$-cs' - 6ss' + s''' = 0. \tag{19}$$

Integrating this gives

$$-cs - 3s^2 + s'' = 0; \tag{19'}$$

multiplication by $2s'$ and integration gives

$$-cs^2 - 2s^3 + s'^2 = 0. \tag{19''}$$

The explicit solution of (19)″ is

$$s = -\frac{c}{2}\mathrm{sech}^2 \frac{\sqrt{c}}{2}x. \tag{20}$$

Using (20), or (19)′ and (19)″, we can verify that

$$\mathcal{L}(s)w = -\frac{c}{4}w, \tag{21}$$

where $\mathcal{L}(u) = -\partial_x^2 + u$ the operator defined in (1), and the eigenfunction w is

$$w = \sqrt{s}.$$

We have seen in Chapter II that the quantities η_j occurring in formula (27)′ are related to the eigenvalues λ of $\mathcal{L}$ by

$$\lambda_j = -\eta_j^2.$$

According to (21), the eigenvalues of $\mathcal{L}$ are related to the soliton speeds c by

$$\lambda_j = -\frac{c_j}{4}. \tag{22}$$

Combining the two gives

$$c_j = 4\eta_j^2. \tag{22'}$$

Set $x = c_j t + z$ into (17) and use (22)′:

$$g_i = C_i(0)e^{-z}. \tag{23}$$

In case i), $|x - c_j t| = |z| > vt$. Therefore for t large, g_i is either very large or very small. It follows from formula (18) that the value of det M_S, given by (18), is largest when S includes all indices i for which $x - c_j t < 0$ and none for which $x - c_j t > 0$:

$$\max_S \det M_S = h_K \left(\prod_{K<i} g_i \right)^2, \tag{24}$$

where we have arranged the η_j in increasing order, and

$$x - c_K t > 0, \qquad x - c_{K+1} t < 0. \tag{24'}$$

As t tends to ∞, the largest term (24) is very much larger than all the other terms in (16); so asymptotically $\det(I + M)$, given by formula (16), is

$$D = \det(I + M) \simeq h_K \left(\prod_{K<i} g_i \right)^2. \tag{25}$$

Taking logarithms and using formula (17) again we get

$$\log D \simeq \log h_K + 2 \sum_{K<i} \log g_i$$

$$= \log h_K + 2 \sum_{K<i} \log C_i(0) + 2 \left(\sum_{K<i} \eta_i^3 \right)t - 2 \left(\sum_{K<i} \eta_i \right)x. \tag{26}$$

In words: in the indicated range of x and t, $logD$ is a linear function of x. According to formula (30)′ of Chapter II,

$$u(x,t) = -2\frac{d^2}{dx^2}\log D(x,t). \tag{27}$$

The same argument that showed that $D(x,t)$ is asymptotically linear shows that its second derivative is asymptotically zero. This proves that for large t, $u(x,t) \simeq 0$ at all points x that are far away from $c_j t$, $j = 1, \ldots, N$.

We turn now to those points x that are near $c_K t$. For such values of x, and t large, *two* of the determinants $\det M_S$ in (18) are very much larger than all the others; these two are $S_0 = \{K, \ldots, N\}$ and $S_1 = \{K+1, \ldots, N\}$. The value of these two determinants are related by

$$\det M_0 = a(e^{4\eta_K^3 t - \eta_K x})^2 \det M_1,$$

a some constant independent of x and t. The determinant D, given by formula (16), is asymptotically the sum of these two terms:

$$D \simeq \det M_0 + \det M_1 = \\ [ae^{8\eta_K^3 t - 2\eta_K x} + 1] \det M_1.$$

Taking logarithms we get

$$logD \simeq \log[ae^{8\eta_K^3 t - 2\eta_K x} + 1] + \log \det M_1. \tag{28}$$

It can be shown, as before, that this asymptotic relation can be differentiated twice with respect to x. We saw before that $\log \det M_1$ is linear in x, so its second derivative is zero. In the first term we set

$$x = c_K t + z.$$

By (22)′, $c_K = 4\eta_K^2$; so we get

$$\frac{d^2}{dx^2}\log D \simeq \frac{d^2}{dz^2}\log[ae^{-2\eta_K z} + 1]. \tag{28′}$$

By a phase shift in z we can make the constant $a = 1$. Carrying out the differentiation we get

$$\eta^2 \operatorname{sech}^2 \frac{\sqrt{c}}{2} z.$$

Setting this in (27) and using $4\eta^2 = c$ we get

$$u(x,t) \simeq -\frac{c}{2}\operatorname{sech}^2 \frac{\sqrt{c}}{2}(x - ct + \varphi)$$

when $|x - ct|$ is bounded. Comparing this with formula (20) we conclude that in the indicated range $u(x,t)$ is for large $|t|$ asymptotically shaped like a soliton.

When the initial potential is not reflectionless, it is possible to show that for the ranges investigated above, formula (15) is a good approximation to the Fredholm determinant, and therefore the asymptotic description of $u(x,t)$ as a soliton train is valid for x and t both large positive, or both large negative. For t large positive and x large negative is is a different story; we shall not go into it.

We conclude this discussion by the following observation:

Theorem 1: *Let $\mathcal{L}$ be the operator $-\partial^2 + u$ on $\mathbb{R}$, where the potential u dies down rapidly as $x \to \pm\infty$. Suppose $\mathcal{L}$ has N eigenvalues $\lambda_1, \ldots, \lambda_N$. Then*

$$\frac{16}{3} \sum \lambda_j^{3/2} \leq \int u_-^2(x)\, dx, \tag{29}$$

where $u_- = \min(u, 0)$ is the negative part of u.

Proof: We have seen in Chapter II that

$$H^{(0)}(u) = \int u^2\, dx$$

is a conserved quantity under the KdV flow. As we have indicated, for t large positive and x large positive, $u(x, t)$ is very nearly the superposition of solitons:

$$u(x, t) \simeq \sum_1^N s(x - c_j t - \varphi_j, c_j).$$

As $t \to \infty$, these solitons move apart; therefore

$$\int\limits_{vt}^{\infty} u^2(x, t)\, dx \simeq \sum \int\limits_{-\infty}^{\infty} s^2(x, c_j)\, dx,$$

and so for large t

$$\int\limits_{-\infty}^{\infty} u^2(x, t)\, dx \geq \sum \int s^2(x, c_j) + \epsilon(t). \tag{30}$$

Using formula (20) we evaluate

$$\int s^2(x, c)\, dx = \frac{c^2}{4} \int \operatorname{sech}^4 \frac{\sqrt{c}}{2} x\, dx = \frac{c^{3/2}}{2} \int \operatorname{sech}^4 y\, dy =$$

$$= 8c^{3/2} \int\limits_0^{\infty} \left(\frac{1}{z + z^-}\right)^4 \frac{dz}{z} = \frac{2}{3} c^{3/2}.$$

Using (22) to express c as 4λ we obtain

$$\int u^2(x, t) \geq \frac{16}{3} \sum \lambda_j^{3/2} + \epsilon(t).$$

Since $\int u^2\, dx$ is invariant, we can replace the left side above by $\int u^2(x)\, dx$, where $u(x)$ is the initial value of u, and can be taken as an arbitrary rapidly decreasing function on $\mathbb{R}$. Finally we note that, it follows from the min-max principle for eigenvalues that the eigenvalues of $\mathcal{L}(u_-)$ are smaller, i.e. larger negative, than the eigenvalues of $\mathcal{L}(u)$; this completes the proof of (29).

The reflectionless solutions given by formulas (27) and (15) have an analogue for solutions of the KdV that are periodic in x. These solutions have the property that the spectrum of the operator $\mathcal{L}(u)$, that in the periodic case is discrete, has multiplicity 2, except for a finite number of eigenvalues. These solutions can be characterized variationally; Krichever and Tanaka have found expressions for them in terms of Riemann θ-functions, described in Toda's book.

102

References

[1] Arnold, V.I., "Mathematical Methods of Classical Mechanics", Grad Texts in Math., 1978, Springer Verlag.

[2] Deift, P. and Trubowitz, E., "Inverse scattering on the line", CPAM **32**, 1979, p. 121–251.

[3] Faddeev, L.D. and Zakharov, V.E., "Korteweg-de Vries equation: a completely integrable Hamiltonian system", Functional Anal. and its Appl. **5**, 1972. p. 280–287.

[4] Fokas, A.S. and Zakharov, V.E., "Important Developments in Soliton Theory", Springer Series in Nonlinear Dynamics, 1993, Springer Verlag.

[5] Gardner, C.S., Greene, J.M., Kruskal, M.D. and Miura, R., "Method for solving the Korteweg-de Vries equation", Phys. Rev. Lett. **19**, 1967, p. 1095–1097.

[6] Kay, I. and Moses, H.E., "Reflectionless transmission through dielectrics and scattering potentials", J. Appl. Physics **27**, 1956, p. 1503–1508.

[7] Kruskal, M.D. and Zabusky, N.J., "Interaction of solitons in a collisionless plasma and the recurrence of initial states", Phys. Rev. Letters **15**, 1965, p. 240–243.

[8] Lamb, G.L. Jr., "Elements of soliton theory", 1980, Pure and Appl. Mathematics Series, Wiley-Interscience.

[9] Lax, P.D., "Integrals of nonlinear equations of evolution and solitary waves", CPAM **21**, 1968, p. 467–490.

[10] Lax, P.D., "Hyperbolic systems of conservation laws and the mathematical theory of shock waves", 1973, SIAM.

[11] Lax, P.D., "Periodic solutions of the KdV equation", CPAM **28**, 1975, p. 141–188.

[12] Lax, P.D., "Almost periodic solutions of the KdV equation", SIAM Rev. **18**, p. 351–375.

[13] Lax, P.D., "Trace formulas for the Schroedinger operator", CPAM **48**, 1994, p. 503-512.

[14] Lax, P.D. and Levermore, C.D., "The small dispersion limit of the KdV equation", CPAM **36**, 1983, p. 253–290, 571-593, 809–830.

[15] Mc Kean, H. and van Moerbeke, P., "The spectrum of Hill's equation", Inv. Math. **30**, 1975, p. 217–274.

[16] Moser, J., "Integrable Hamiltonian Systems and Spectral Theory", Acc. Naz. Lincei, Scuola Normale Superiore, Lezioni Fermiane, Pisa, 1981.

[17] Toda, M., "Theory of Nonlinear Lattices", 2nd Ed., 1988, Springer Series in Solid-State Sciences, Springer Verlag.

Nonlinear Hyperbolic-Dissipative Partial Differential Equations

Tai-Ping Liu
Department of Mathematics
Stanford University

1. Introduction

Consider the general system of nonlinear evolutionary partial differential equations of conservation form

$$u_t + f(u)_x = (B(u)u_x)_x + g(u,x,t). \tag{1.1}$$

Here $u = u(x,t)\epsilon R^n$ represents the density , $f(u)$ the flux, and $(B(u)u_x)_x$ the viscosity of physical variables. The term $g(u,x,t)$ may represent reaction, damping, or relaxation effects. The system is usually quasilinear. The simplest system is the hyperbolic conservation laws

$$u_t + f(u)_x = 0. \tag{1.2}$$

It possesses shock waves, simple waves and linear hyperbolic waves, cf. Dafermos' note in the series. As usual, we assume that the system is hyperbolic, that is, the flux matric $f'(u)$ has real eigenvalues $\lambda_1(u) \leq \lambda_2(u) \leq \cdots \leq \lambda_n(u)$,

$$f'(u)r_i(u) = \lambda_i(u)r_i(u), \quad l_i(u)f'(u) = \lambda_i(u)l_i(u), \quad l_i(u) \cdot r_j(u) = \delta_{ij}$$

$$i,j = 1, \cdots, n. \tag{1.3}$$

An important physical example is the compressible Euler equations.

Another important class of physical systems is the viscous conservation laws

$$u_t + f(u)_x = (B(u)u_x)_x. \tag{1.4}$$

Physical viscosity $B(u)$ is usually not uniform and so (1.4) is then quasilinear and not uniformly parabolic, but hyperbolic-parabolic. To understand some of the basic phenomena one also consider the semilinear system with artificial viscosity

$$u_t + f(u)_x = \varepsilon u_{xx}. \tag{1.5}$$

Here ε represents the strength of the viscosity. Although the solutions of the viscous conservation laws are expected to approach those of the hyperbolic conservation laws as the viscosity tends to zero, this zero dissipation limit is quite complicated. It may happen, for instance, some viscous waves may become non-physical, i.e. unstable, as the viscosity tends to zero as in the case of MHD and nonlinear elasticity. Also, different viscosity matrix may give rise to different inviscid waves as in the case of multipahse flows and in combustions. Several basic physical models are of the form (1.4), e.g. the compressible Navier-Stokes equations, viscous elasticity and magneto-hydrodynamics (MHD) equations.

The term $g(u, x, t)$ can also induce dissipation. An important class is the hyperbolic conservation laws with relaxation

$$v_t + f(v, w)_x = 0$$

$$w_t + g(v, w)_x = h(v, w). \tag{1.6}$$

Here the vector v represents the conserved pysical quantities. The function $h(v, w)$ often takes the form

$$h(v, w) = \frac{W(v) - w}{\varepsilon}, \tag{1.7}$$

with $W(v)$ as the equilibrium value for v and $\varepsilon > 0$ the relaxation time. The system is assumed to be strongly coupled and with variable equilibrium states, that is, the function $f(v, w)$ depends strongly on the second vector w and that $W(v)$ depends strongly on v. This allows for rich wave phenomena as the result of partial disspation induced by the relaxation. Physical examples includes the kinetic models, gas in thermo-non-equilibrium and elasticity with fading memory.

The remaining effects are the reaction and damping, depending on whether the source $g(u, x, t)$ is an increasing or decreasing function of u. We will discuss these in the last sections. An interesting aspect of these effects is that they are usually only partial and not uniform. Physical examples include the porous media flows, combustions and moving sources.

2. Scalar Viscous Conservation Laws

Consider the scalar viscous conservation laws

$$u_t + f(u)_x = u_{xx}, \tag{2.1}$$

where, for simplicity, we have set the viscosity coefficient to be one. The simplest example is the Burgers equation

$$u_t + \left(\frac{u^2}{2}\right)_x = u_{xx}. \tag{2.2}$$

It can be solved by the well-known Hopf-Cole transformation, [Whitham]. Because of the scaling property the Burgers equation possesses explicit solutions which correspond to the heat kernel of the heat equation:

$$\theta(x, t; c) \equiv (t + 1)^{-1/2} \xi\left(\frac{x}{\sqrt{t+1}}; c\right),$$

$$\xi(y; c) \equiv \frac{(e^{c/2} - 1)e^{-y^2/4}}{\sqrt{\pi} + (e^{c/2} - 1) \int_{y/2}^{\infty} e^{-\tau^2} d\tau},$$

$$\theta(x, 0; c) = c\delta(x). \tag{2.3}$$

The travelling waves ,the viscous shock waves, of the Burgers equation are

$$\phi(x) \equiv u_0 \frac{1 - e^{u_0 x}}{1 + e^{u_0 x}}, \tag{2.4}$$

where, for simplicity, we only consider the stationary shocks connecting positive states u_0 at $x = -\infty$ and $-u_0$ at $x = \infty$. There is another important class of waves, the expansion, or the rarefaction, waves. For Burgers equation we have the following family of explicit solutions

$$\psi(x,t) \equiv (-\ln \zeta(x,t))_x, \tag{2.5}$$

$$\zeta(x,t) \equiv e^{u_0 x + u_0^2 t/2} \int_{-\infty}^{0} G(x + u_0 t - \tau, t) d\tau$$

$$+ e^{-u_0 x + u_0^2 t/2} \int_{0}^{\infty} G(x - u_0 t - \tau, t) d\tau,$$

$$\psi(x,0) = \begin{cases} -u_0, & \text{for } x < 0, \\ u_0 & \text{for } x > 0. \end{cases}$$

Here again we have chosen only rarefaction waves symmetrically situated around $x = 0$. The function G is the linear heat kernel

$$G(x,t) \equiv \frac{1}{\sqrt{4\pi t}} e^{-\frac{x^2}{4t}}. \tag{2.6}$$

The above waves are nonlinearly stable, though in different senses. We study their stability in the remaining of this section.

2.1 Nonlinear Diffusion Waves

We are interested in the time-asymptotic behaviour of the solutions of (2.1) which are perturbation of constant states, taken to be zero here. To have nonlinear behavior we assume that the function $f(u)$ is convex at $u = 0$. By a change of the dependent variable we may assume that

$$f''(0) = 1, \quad f'(0) = 0. \tag{2.7}$$

The initial value is assumed to have finite total mass

$$\int_{-\infty}^{\infty} u(x,0) dx = c. \tag{2.8}$$

We will show that the solution tends to the self-similar solutions $\theta(x,t;c)$, (2.3) of the Burgers equation. We have from (2.1) for $u(x,t)$, (2.2) for $\theta(x,t)$ and (2.3), (2.7) and (2.8) that

$$v_t = v_{xx} + [-\theta v - \frac{v^2}{2} + O(1)(|v|^3 + |\theta|^3)]_x,$$

$$\int_{-\infty}^{\infty} v(y,t) dy = 0, \quad t \geq 0, \quad v \equiv u - \theta. \tag{2.9}$$

By the Duhamel's principle

$$v(x,t) = \int_{-\infty}^{\infty} G(x - y, t)v(y, 0)dy$$

$$+ \int_0^t \int_{-\infty}^{\infty} [-\theta v - \frac{v^2}{2} + O(1)(|v|^3 + |\theta|^3)]_y(y, s)G(x - y, t - s)dyds$$

$$= -\int_{-\infty}^{\infty} G_y(x - y, t)[\int_{-\infty}^{y} v(z, 0)dz]dy$$

$$+ \int_0^t \int_{-\infty}^{\infty} O(1)[\theta v + \frac{v^2}{2} + O(1)(|v|^3 + |\theta|^3)](y, s)G_y(x - y, t - s)dyds. \quad (2.10)$$

To make the analysis simple, we assume that the diffusion wave and its perturbation are weak:

$$|c| + \sup_{-\infty < x, \infty} |v(x, 0)|(1 + x^2)^{3/4} \equiv \varepsilon, \quad (2.11)$$

for small ε. We will show that

$$v(x, t) = O(1)\varepsilon(x^2 + t + 1)^{-3/4}, \quad -\infty < x < \infty, \quad t \geq 1. \quad (2.12)$$

First we see that, with (2.11), the first integral of (2.10) is, for M sufficiently large,

$$\int_{-\infty}^{\infty} O(1)\frac{x-y}{\sqrt{t}}t^{-1}e^{-\frac{(x-y)^2}{4t}}\varepsilon(1 + y^2)^{-1/4}dy = \int_{|y|>M\sqrt{t}} O(1)t^{-1}e^{-\frac{(x-y)^2}{4t}}t^{-1/2}dy$$

$$+ \int_{|y|<M\sqrt{t}} O(1)t^{-1}\varepsilon(y^2 + 1)^{-1/4}dy$$

$$= O(1)\varepsilon t^{-3/4}, \quad \text{for } |x| < M\sqrt{t}/2;$$

$$\int_{-\infty}^{\infty} O(1)\varepsilon t^{-1/2}e^{-\frac{(x-y)^2}{4t}}(y^2 + 1)^{-3/4}dy = [\sup_{-\infty<y<\infty} e^{-\frac{(x-y)^2}{8t}}(y^2 + 1)^{-3/4}]$$

$$\cdot \int_{-\infty}^{\infty} O(1)\varepsilon t^{-1/2}e^{-\frac{(x-y)^2}{8t}}dy$$

$$= O(1)\varepsilon(x^2 + 1)^{-3/4}, \quad \text{for } |x| > M\sqrt{t}.$$

Next we show that, with the apriori hypothesis (2.12), the second integral

$$\int_0^t \int_{-\infty}^{\infty} O(1)\varepsilon^2[(s + 1)^{-5/4}e^{-\frac{y^2}{4(s+1)}} + (y^2 + s + 1)^{-3/2}](t-s)^{-1}e^{-\frac{(x-y)^2}{4(t-s)}}\frac{x-y}{\sqrt{t-s}}dyds$$

is much smaller than the left-hand side of (2.11). We estimate this by considering the following:

$$i \equiv \int_0^t \int_{-\infty}^{\infty} (s + 1)^{-5/4}e^{-\frac{y^2}{4(s+1)}}(t - s)^{-1}e^{-\frac{(x-y)^2}{4(t-s)}}dyds$$

$$= \int_0^t O(1)(t - s)^{-1/2}(s + 1)^{-3/4}(t + 1)^{-1/2}e^{-\frac{x^2}{4(t+1)}}ds$$

$$= O(1)[\int_0^{t/2} O(1)(t + 1)^{-1/2}t^{-1/2}(s + 1)^{-3/4}ds$$

$$+ \int_{t/2}^{t} (t+1)^{-5/4}(t-s)^{-1/2}ds]e^{-\frac{x^2}{4(t+1)}},$$

$$ii \equiv \int_0^t \int_{-\infty}^{\infty} (y^2 + s + 1)^{-3/2}(t-s)^{-1}e^{-\frac{(x-y)^2}{4(t-s)}}dyds$$

$$= \int_0^t \int_{-\infty}^{-\sqrt{s+1}} O(1)|y|^{-3}(t-s)^{-1}e^{-\frac{(x-y)^2}{4(t-s)}}dyds$$

$$+ \int_0^t \int_{-\sqrt{s+1}}^{\sqrt{s+1}} O(1)(s+1)^{-3/2}(t-s)^{-1}e^{-\frac{(x-y)^2}{4(t-s)}}dyds$$

$$+ \int_0^t \int_{-\sqrt{s+1}}^{\infty} O(1)y^{-3}(t-s)^{-1}e^{-\frac{(x-y)^2}{4(t-s)}}dyds$$

$$\equiv ii_1 + ii_2 + ii_3.$$

The term ii_2 is dominated by the first integral in (2.10). We consider the case $x > 0$, for which ii_1 is dominated by ii_3. When $0 \le x \le \sqrt{t+1}$

$$ii_3 = O(1)[\int_0^{t/2} \int_{\sqrt{s+1}}^{\infty} t^{-1}y^{-3}dyds$$

$$+ \int_{t/2}^{t} \int_{\sqrt{s+1}}^{\infty} (t-s)^{-1}(s+1)^{-3/2}e^{-\frac{(x-y)^2}{4(t-s)}}dyds$$

$$- O(1)[\int_0^{t/2} t^{-1}(s+1)^{-1}ds + \int_{t/2}^{t} (t-s)^{-1/2}(s+1)^{-3/2}ds]$$

$$= O(1)[t^{-1}\ln(t+1) + (t+1)^{-1}].$$

For $x \ge \sqrt{t+1}$

$$ii_3 = O(1)[\int_0^t \int_{\sqrt{s+1}}^{Mx} y^{-3}(t-s)^{-1}e^{-\frac{x^2}{4(t-s)}}dyds$$

$$+ \int_0^t \int_{Mx}^{\infty} (x+1)^{-3}(t-s)^{-1}e^{-\frac{(x-y)^2}{4(t-s)}}dyds]$$

$$= O(1)[\int_0^t ((s+1)^{-1}e^{-\frac{x^2}{4(t-s)}} + (x+1)^{-3}e^{-\frac{(x-y)^2}{4(t-s)}})(t-s)^{-1}ds]$$

$$= O(1)[t^{-1}\ln(t+1) + (x+1)^{-3}\ln t]e^{-\frac{x^2}{Dt}},$$

where the constant D is close to 4 when M is chosen large. The final case of $x < 0$ is similar. We have thus proved the estimate (2.12). It follows that

$$|v|_{L_p(x)} = O(1)t^{(-3p+2)/(4p)}, \quad \text{for } t > 0. \tag{2.13}$$

The decay rates in (2.12) and (2.13) can be improved when the initial value decays faster than $(x^2 + 1)^{-3/2}$. The optimal rates are

$$|v|_{L_p(x)} = O(1)t^{-1+1/(2p)}.$$

The rate (2.13), though, is optimal for system due to nonlinear interactions of waves of different characteristic families, as we will see in Section 3.1.

2.2. Visccous Shock Waves

We now study the stability of viscous shock waves. For the inviscid hyperbolic conservation law (1.2) the shock is stable because it is compressive

$$f'(u_-) > s > f'(u_+),$$

where u_- (u_+) are the states to the left (right) of the shock and s the shock speed. This implies that a perturbation, which propagates with characteristic speed $f'(u)$, will be absorbed into the shock and thus results in the stability of the shock. For the viscous conservation laws there is the additional effect of diffusion. While the perturbation still propagates along the general direction of the characteristics, it also diffuses around it. To see the exact process of the balance of the nonlinear effect of compression and the effect of diffusion we consider the Burgers equation. For simplicity, we consider a perturbation of the stationary shock (2.4)

$$u(x,0) = \phi(x) + v(x,0). \tag{2.14}$$

We may translate the shock so that the perturbation has zero integral. Since both $u(x,t)$ and $\phi(x)$ satisfy the conservation law (2.2) we have

$$\int_{-\infty}^{\infty} v(y,t)dy = 0, \quad \text{for } t \geq 0, \tag{2.15}$$

$$v_t + (\phi v)_x = v_{xx} - (\frac{v^2}{2})_x. \tag{2.16}$$

There are several ways of studying the decay of the perturbation, i.e. the asymptotic stability of the shock. We present here the analysis based on the Green function of the linearized operator in (2.16). The Green function is easier to find if the equation is integrated

$$w_t + \phi w_x = w_{xx} - \frac{v^2}{2}.$$

Multiply this by

$$e^{-\frac{1}{2} \int^x \phi(y)dy}$$

and use the fact that ϕ is a stationary wave of the Burgers equation we have

$$z_t + \frac{1}{4}u_0^2 z = z_{xx} - \frac{1}{2}v^2 e^{-\frac{1}{2} \int^x \phi(y)dy}.$$

The linear part of this equaiton has explicit Green function and we have from the Duhamel's principle that

$$e^{u_0^2 t/4} z(x,t) = \int_{-\infty}^{\infty} \frac{1}{\sqrt{4\pi t}} e^{-\frac{(x-y)^2}{4t}} z(y,0)dy - \int_0^t \int_{-\infty}^{\infty} e^{\frac{1}{2}\int^y \phi} e^{u_0^2 s/4} \frac{v^2}{2}(y,s)dyds.$$

In terms of the function $w(x,t)$ this becomes

$$w(x,t) = \int_{-\infty}^{\infty} \frac{1}{\sqrt{4\pi t}} \frac{1 + e^{u_0 y}}{1 + e^{u_0 x}} e^{-\frac{(y-x+u_0 t)^2}{4t}} dyw(y,0)$$

$$-\int_0^t \int_{-\infty}^\infty \frac{1}{\sqrt{4\pi(t-s)}} \frac{1+e^{u_0 y}}{1+e^{u_0 x}} e^{-\frac{(y-x+u_0(t-s))^2}{4(t-s)}} \frac{v^2}{2}(y,s)\,dy\,ds$$

$$= \int_{-\infty}^\infty \frac{1}{\sqrt{4\pi t}} \frac{1+e^{-u_0 y}}{1+e^{-u_0 x}} e^{-\frac{(y-x-u_0 t)^2}{4t}} w(y,0)\,dy$$

$$-\int_0^t \int_{-\infty}^\infty \frac{1}{\sqrt{4\pi(t-s)}} \frac{1+e^{-u_0 y}}{1+e^{-u_0 x}} e^{-\frac{(y-x-u_0(t-s))^2}{4(t-s)}} \frac{v^2}{2}(y,s)\,dy\,ds, \tag{2.17}$$

whence we can also derive equation for $v(x,t)$ by differentiation with respect to x.

Notice that the propagation of the initial data $w(y,0)$ is along the characteristic direction u_+. However, it is weighted by the factor

$$\frac{1+e^{u_0 y}}{1+e^{u_0 x}} \quad \text{or} \quad \frac{1+e^{-u_0 y}}{1+e^{-u_0 x}}.$$

For instance, when $x > 0$ we may take the second expression in (2.17) and the propagation of the initial data $w(y,0)$, $y > 0$, is similar to the linear heat equaion and with speed $-u_0$. That is, it is similar to the inviscid case with straightforward addition of diffusion. On the other hand, for the propagation of the initial data to the left, $w(y,0)$, $y < 0$, we take the first expression in (2.17) and see that the propagation, though with speed u_0 as for the inviscid case, does not stop at the shock, but continue to the right of the shock at (x,t). This diffusion effect diminishes as (x,t) moves away from the shock at the rate $e^{-u_0 x}$. This has the consequence that the perturbation decays at the same rate as the decay rate of the initial perturbation at $x = \pm\infty$.

We now use the above observation to study the pointwise decay of $v(x,t)$ based on the following smallness of the initial value and the apriori hypothesis:

$$\sup_{-\infty<x<\infty} [|v(x,0)|(x^2+1)^{\alpha/2} + |w(x,0)|(x^2+1)^{(\alpha-1)/2}] \equiv \varepsilon, \quad \alpha > 1, \tag{2.18}$$

$$w(x,t) = O(1)\varepsilon[(|x|+u_0 t)^2 + 1]^{(-\alpha+1)/2}, \tag{2.19}_1$$

$$v(x,t) = O(1)\varepsilon[(|x|+u_0 t)^2 + 1]^{-\alpha/2}. \tag{2.19}_2$$

We now prove (2.19) using (2.17) under the assumption that ε is small. For definiteness, set $x > 0$ and use the following combination of the expresion of (2.17)

$$w(x,t) = \int_{-\infty}^0 \frac{1}{\sqrt{4\pi t}} \frac{1+e^{u_0 y}}{1+e^{u_0 x}} e^{-\frac{(y-x+u_0 t)^2}{4t}} w(y,0)\,dy$$

$$+ \int_0^\infty \frac{1}{\sqrt{4\pi t}} \frac{1+e^{-u_0 y}}{1+e^{-u_0 x}} e^{-\frac{(y-x-u_0 t)^2}{4t}} w(y,0)\,dy$$

$$- \int_0^t \int_{-\infty}^0 \frac{1}{\sqrt{4\pi(t-s)}} \frac{1+e^{u_0 y}}{1+e^{u_0 x}} e^{-\frac{(y-x+u_0(t-s))^2}{4(t-s)}} v^2(y,s)\,dy\,ds$$

$$- \int_0^t \int_0^\infty \frac{1}{\sqrt{4\pi(t-s)}} \frac{1+e^{-u_0 y}}{1+e^{-u_0 x}} e^{-\frac{(y-x-u_0(t-s))^2}{4(t-s)}} v^2(y,s)\,dy\,ds$$

$$= O(1)[\int_{-\infty}^{0} t^{-1/2} e^{-u_0 x} e^{-\frac{(y-x+u_0 t)^2}{4t}} w(y,0)dy$$

$$+ \int_{0}^{\infty} t^{-1/2} e^{-\frac{(y-x-u_0 t)^2}{4t}} w(y,0)dy$$

$$+ \int_{0}^{t} \int_{-\infty}^{0} (t-s)^{-1/2} e^{-u_0 x} e^{-\frac{(y-x+u_0(t-s))^2}{4(t-s)}} v^2(y,s)dyds$$

$$+ \int_{0}^{t} \int_{0}^{\infty} (t-s)^{-1/2} e^{-\frac{(y-x-u_0(t-s))^2}{4(t-s)}} v^2(y,s)dyds]$$

$$\equiv O(1)[i + ii + iii + iv].$$

The terms $i \sim iv$ are estimated using (2.19). We need to show that i and ii are of the form of $(2.19)_1$ and that iii and iv are smaller than the expression in $(2.19)_1$. We note here only that the decays (2.19) is slow when the shock strengh $2u_0$ is small. This is so because in that case the information reaching the point (x,t) comes from sources almost directly below the point; while for strong shocks, the information comes from further out and decays at faster rates, (2.19).

It should be mentioned that the energy method is always important for the study of nonlinear waves. It provides an easy proof of the nonlinear stability of viscous shocks, though without the pointwise estimates. This is done as follows: Multiply the interagted version of (2.16) by w and integrate to yield

$$\int_{-\infty}^{\infty} w^2(x,t_2)dx + \int_{t_1}^{t_2} \int_{-\infty}^{\infty} [w_x^2(1 + \frac{1}{2}w) + \frac{1}{2}|\phi_x|w^2 dxdt = \int_{-\infty}^{\infty} w^2(x,t_1)dx.$$

With the crucial compression property of the shock $\phi_x < 0$ and the a priori hypothesis that the perturbation w is small, the above provides the basic energy estimate

$$\int_{-\infty}^{\infty} w^2(x,t_2)dx + \int_{t_1}^{t_2} \int_{-\infty}^{\infty} [w_x^2 + |\phi_x|w^2](x,t)dxdt = O(1)\int_{-\infty}^{\infty} w^2(x,t_1)dx,$$

$$0 \leq t_1 \leq t_2.$$

From this we obtain similar energy estimate by integrating (2.16) times v and $(2.16)_x$ times v_x:

$$\int_{-\infty}^{\infty} [v^2 + v_x^2](x,t_2) + \int_{t_1}^{t_2} \int_{-\infty}^{\infty} [v_x^2 + v_{xx}^2]dxdt = V(1)\int_{-\infty}^{\infty} [w^2 + v^2 + v_x^2](x,t_1)dx.$$

From this we have

$$\int_{0}^{\infty} \int_{-\infty}^{\infty} v_x^2 dxdt < \infty,$$

and so

$$\int_{t}^{\infty} \int_{-\infty}^{\infty} v_x^2 dxdt \to 0,$$

as t tends to infinity. This and the above energy estimates yield

$$\frac{1}{2}v^2(x,t) = \int_{-\infty}^{x} vv_x(y,t)dy \leq [\int_{-\infty}^{\infty} v^2(y,t I dy \int_{-\infty}^{\infty} v_x^2(y,t)dy]^{-1/2}$$

$$O(1)[\int_{-\infty}^{\infty}[w^2+v^2](y,0)dy]^{-1/2}[\int_{-\infty}^{\infty}[w^2+v^2+v_x{}^2](y,0)dy+\int_{t-1}^{t}\int_{-\infty}^{\infty}v_x{}^2(y,s)dyds]$$

which tends to zero as t tends to infinity. The energy method with weighted norm is an effective method for the scalar equation to obtain the stability result with rates of convergence. The method is also effective in studying the stability of strong shocks for equation with nonconvex flux, [Matsumura-Nishihara]. For systems to be discussed latter, the exact method of pointwise estimates using the Green functions has been fruitful.

2.3 Viscous Rarefaction Waves

Rafaction waves are waves which expand, i.e. the characteristic value is an increasing function of x

$$f'(u)_x(x,t) \geq 0.$$

Consider first the particular rarefaction wave for the Burgers equation (2.2) with the jump data, the Riemann data,

$$u(x,0) = \begin{cases} -u_0, & x < 0, \\ u_0, & x > 0, \end{cases} \qquad (2.20)_1$$

$$u_0 > 0.$$

Here we have taken, for simplicity, the rarefaction wave symmetric about the t-axis. The problem can be solved easily using the well-known Hopf-Cole transformation and the solution is

$$U(x,t) = u_0\frac{\phi_+(x,t)-\phi_-(x,t)}{\phi_+(x,t)+\phi_-(x,t)},$$

$$\phi_+(x,t) = \frac{1}{\sqrt{4\pi t}}e^{-2u_0x+u_0{}^2t}\int_0^{\infty}e^{-\frac{(2x-2u_0t-\eta)^2}{4t}}d\eta,$$

$$\phi_-(x,t) \equiv \frac{1}{\sqrt{4\pi t}}e^{2u_0x+u_0{}^2t}\int_{-\infty}^0 e^{-\frac{(2x+2u_0t-\eta)^2}{4t}}d\eta. \qquad (2.20)_2$$

It turns out that general solutions of the Burgers equation with initial data whose end state at $x = -\infty$ less than that at $x = \infty$ converge to particular solutions as (2.20). We now outline the proof of such a theorem.

Given a perturbation of $U(x,t)$

$$u(x,0) = U(x,0) + v(x,0).$$

From the Burgers equation

$$v_t + (Uv)_x = v_{xx} - (\frac{v^2}{2})_x,$$

$$w_t + Uw_x = w_{xx} - \frac{v^2}{2}, \qquad w_x \equiv v. \qquad (2.21)$$

In the above we have properly translated the function $U(x,t)$ so that

$$\int_{-\infty}^{\infty}v(y,t)dy = 0, \qquad t \geq 0.$$

Multiply (2.20) by

$$a(x,t) \equiv e^{-\frac{1}{2}\int_{-\infty}^{x}(U(\xi,t)+u_0)d\xi}$$

to obtain the following constant coefficients linear part

$$z_t - u_0 z_x = z_{xx} + \frac{1}{a(x,t)}\frac{v^2}{2}, \quad z(x,t) \equiv w(x,t)a(x,t).$$

This equation can be solved by the Duhamel's principle since the Green function for the linear part is simply

$$G(x,t) \equiv \frac{1}{\sqrt{4\pi t}}e^{-\frac{(x-u_0t)^2}{4t}}.$$

Thus we have from the above identities that

$$w(x,t) = \int_0^\infty G(x-y,t)\frac{a(x,t)}{a(y,0)}w(y,0)dy$$

$$+ \int_0^t \int_{-\infty}^\infty G(x-y,t-s)\frac{a(x,t)}{a(y,s)}\frac{v^2}{2}(y,s)dyds.$$

Following similar approach of the last two sections we can prove that the function $w(x,t)$ is bounded and that the function $v(x,t)$ has the following decay rates

$$v(x,t) = O(1)\frac{1}{t}, \quad \text{for} \; -u_0t - \sqrt{t} < x < u_0t,$$

and the decay rates for the region $|x| > u_0t + \sqrt{t}$ depends on the decay in x of the initial data as in the case for the viscous shocks.

There is a basic difference between the rarefaction waves and other two types of waves discussed before. Rarefaction waves are stable in $L_p(x)$ norms for any $p > 1$, but not for $p = 1$. This is due to the fact that one may translate parts of the wave away from each other. The diffusion will smooth out the resulting nonsmoothness but not to bring them together. This is a nonlinear hyperbolic phenomenon and the dissipation has a minor effect in this. Because of this, the devise of considering the antiderivative $w(x,t)$ of the perturbation $v(x,t)$ is an technical devise and is not as crucial as in the shock case.

The expansiveness of the rarefaction waves, $U_x > 0$, makes it easier than the shock waves to use the energy method. In fact the basic energy estimate follows readily from integrating the equation for v preceeding (2.21) times v:

$$\int_{-\infty}^\infty \frac{1}{2}v^2(x,t_2)dx + \int_{t_1}^{t_2}\int_{-\infty}^\infty [v_x^2 + |U|^2]dxdt = O(1)\int_{-\infty}^\infty v^2(x,t_1)dx, \quad t_1 < t_2.$$

Standard energy method as in the last section yields the stability of the rarefaction waves without the rates of convergence.

3. System of Viscous Conservation Laws, I

We consider viscous conservation laws (1.4). There are two main difficulties in generalizing the stablity analyses of the last section for the scalar conservation law to the systems. The first is the nonlinear coupling of waves pertaining to the characteristic fields λ_i, $i = 1 \cdots n$, which gives rise to rich wave phenomena. The second difficulty is due to the degenaracy of the viscousity matrix $B(u)$ in physical systems, which makes the system hyperbolic-parabolic and not parabolic, and possessing discontinuous solutions. In this section we study the first difficulty by considering the simplest situation of the nonlinear interaction of diffusion waves. Again our approach is to obtain pointwise estimates through Green functions. Consider the viscous conservation laws with initial value, which is a perturbation of the zero state

$$u_t + f(u)_x = u_{xx}, \tag{3.1}$$

$$u(x, 0) = O(1)\varepsilon(x^2 + 1)^{-3/4}, \tag{3.2}$$

for a small constant ε. The specific rate of decay for the initial value is the minimum one for a simple expression of the large-time behaviour of the solution. As we will see later, due to nonlinear coupling of waves, the decay rate of the solution minus certain linear and nonlinear heat kernel does not improve even if the decay rate of the initial value is assumed to be higher. This is in sharpe contrast to the scalar case in Section 2.1.

3.1 Time-Asymptotic States

The first step is to identify the first-order time-asymptotic state of the solution. This depends on the total mass of the initial value

$$\int_{-\infty}^{\infty} u(x, 0)dx \equiv \sum_{i=1}^{n} c_i r_i(0). \tag{3.3}$$

Here r_i, $i = 1, \cdots n$, are the characteristic direction for the inviscid system, (1.3). Here we make another important simplification throughout Section 3 that the inviscid system is strictly hyperbolic, that is ,

$$\lambda_1(u) < \lambda_2(u) < \cdots < \lambda_n(u). \tag{3.4}$$

At the zero state each characteristic family is either genuinely nonlinear (g.nl.) or linearly degenerate (l.dg.), [Lax 1],

$$\nabla \lambda_i \cdot r_i(0) = 1, \tag{g.nl}$$

$$\nabla \lambda_i \cdot r_i(0) = 0, \tag{l.dg}$$

where we have normalized the eigenvector $r_i(0)$ so that in the (g.nl) case the right hand side is not only nonzero but one. Note that this is consistent with the normalization in (1.3) on the left eigenvectors. With this, the time-asymptotic state turns out to be

$$\Theta(x,t) \equiv \sum_{i=1}^{n} \Theta_i(x,t) \equiv \sum_{i=1}^{n} \theta_i(x,t)r_i(0),$$

$$\theta_i(x,t) \equiv \begin{cases} \theta(x - \lambda_i(0)t, t; c_i) \text{ of } (2.3), & i - \text{field } (g.nl), \\ c_i G(x - \lambda_i(0)(t+1), t+1) \text{ of } (2.6), & i - \text{field } (l.dg). \end{cases} \quad (3.5)$$

In other words, the time-asymptotic state consists of linear heat kernels and nonlinear heat kernels so as to satisfy the zero mass condition

$$\int_{-\infty}^{\infty} v(y,0)dy = 0, \quad t \geq 0,$$

$$v(x,t) \equiv u(x,t) - \Theta(x,t). \quad (3.6)$$

The first identity in (3.6) follows from (3.3) and (3.5), which imply that (3.6) holds at $t = 0$, and from the conservaton laws (3,1) it holds for all $t > 0$. Since the total mass stays constant, the $L_1(x)$ norm of the solution does not decay. Assuming that (3.5) represents an accurate time-asymptotic solution, the zero mass condition (3.6) would imply that $v(x,t)$ decays in $L_1(x)$ and decays faster than $\Theta(x,t)$ in $L_p(x)$, $p > 1$. We will show later through pointwise estimates that this indeed is the case. We first assess the accuracy of (3.5) in the time-asymptotic sense. By Taylor expansion of the fuction $f'(u)$, cf. (1.3), and the conditions $(g.nl)$ and $(l.dg)$

$$\Theta_{it} + f(\Theta_i)_x = \Theta_{ixx} + [\sum_{j \neq i} \frac{1}{2} d_{jii} \theta_i^2 r_j(0) + O(1)|\theta_i|^3]_x, \quad (3.7)$$

$$d_{ijk} \equiv -2l_i f''(r_j, r_k)(0).$$

The specific definition of the diffusion waves (3.5) is to yield the zero mass condition (3.6) and the nonresonance condition $i \neq j$ in the summation in (3.7). Since each Θ_i function propagates in distinct direction due to strict hyperbolicity, (3.4), we have from (3.7) that $\Theta(x,t)$ is an accurate time-asymptotic solution of (3.1):

$$\Theta_t + f(\Theta)_x = \Theta_{xx} + [\sum_{i \neq j} d_{jii} \theta_i^2 r_i(0) + O(1)|\Theta|^3]_x. \quad (3.8)$$

From (3.1), (3.6) and (3.8) we have

$$v_{it} + \lambda_i(0)v_{ix} = v_{ixx} + [-\frac{1}{2}\sum_{i \neq j} d_{jii}\theta_i^2 + \sum_{jk} d_{ijk}\theta_j v_k$$

$$+ O(1)(|v|^2 + |\theta|^3)]_x, \quad (3.9)$$

$$\int_{-\infty}^{\infty} v_i(x,t)dx = 0, \quad t \geq 0, \quad (3.10)$$

$$v(x,t) \equiv \sum_{i=1}^{n} v_i(x,t)r_i(0). \quad (3.11)$$

We will show that

$$v_i(x,t) = O(1)\varepsilon[(x - \lambda_i(0)(t+1)^2 + t + 1]^{-3/4}$$

$$+ \sum_{j \neq i}((x - \lambda_j(0)(t+1))^3 + t^2 + 1)^{-1/2}]. \quad (3.12)$$

This implies that

$$|v|_{L_p(x)} = O(1)\varepsilon(t + 1)^{(-3p+2)/4p}, \qquad (3.13)$$

which is at lower rates than for the scalar case. The rates cannot be improved even if the initial value $u(x,0)$ is of compact support. This is due to the nonlinear coupling of waves of different characteristic families, as we will see in the next section.

3.2 Nonlinear Coupling of Diffusion Waves

With the assumption (3.2), the identity (3.10), the apriori hypothesis (3.12) and the Duhamel's principle for (3.9) we need to consider the following integrals

$$i \equiv \int_{-\infty}^{\infty} \frac{1}{\sqrt{4\pi t}} e^{-\frac{(y-x\lambda_i t)^2}{4t}} (O(1)(y^2 + 1)^{-1/4})_y dy,$$

$$ii \equiv \int_0^t \int_{-\infty}^{\infty} \frac{1}{\sqrt{4\pi(t-s)}} e^{-\frac{(x-y-\lambda_i(t-s))^2}{4(t-s)}} (\theta_j{}^2)_y(y,s) dy ds, \quad j \neq i,$$

$$iii \equiv \int_0^t \int_{-\infty}^{\infty} \frac{1}{\sqrt{4\pi(t-s)}} e^{-\frac{(x-y-\lambda_i(t-s))^2}{4(t-s)}} (\theta_i v_i)_y(y,s) dy ds,$$

$$iv \equiv \int_0^t \int_{-\infty}^{\infty} \frac{1}{\sqrt{4\pi(t-s)}} e^{-\frac{(x-y-\lambda_i(t-s))^2}{4(t-s)}} (O(1)(s + 1)^{-3/2} e^{-\frac{(y-\lambda_j(s+1))^2}{4(s+1)}})_y dy ds,$$

$$v \equiv \int_0^t \int_{-\infty}^{\infty} \frac{1}{\sqrt{4\pi(t-s)}} e^{-\frac{(x-y-\lambda_i(t-s))^2}{4(t-s)}} (O(1)((y - \lambda_j(s+1))^2 + s + 1)^{-3/2})_y dy ds,$$

$$i, j = 1, 2 \cdots, n,$$

where $\lambda_i \equiv \lambda_i(0)$. Note that the term $d_{ijk}\theta_j v_k$, $j, k \neq i$ is absorbed into the terms ii and iv above because $v_k(x,t) = O(1)(t+1)^{-1}$ in the characteristic direction $x = \lambda_j t$. The term $d_{iii}\theta_i v_i$ needs to be treated separately and is listed as iii above.

We need to show that the above integrals decay at rates no faster than those given by (3.12). This involves long and tedious computations. It is this kind of computations that reveals the effects of nonlinear interactions of waves pertaining to different characteristic families. They are therefore essential for the understanding of the behaviour of solutions for the system of viscous conservatrion laws. We will not carry out all the computations, except for the following remarks of qualitative nature. The integrals i, iv and v are estimated by first performing the integration by parts so that the integrants gain an additional factor of $(t - s)^{-1/2}$. One needs to break the integral into several parts according to the relation of (x,t) to the characteristic lines $x - \lambda_j t$, $j = 1, 2, \cdots, n$. We illustrate this by calculating out one of the simplest terms, namely, the term iv above with $i \neq j$. For simplicity we take $\lambda_i = 0$ and $\lambda_j = 1$.

$$iv_1 \equiv \int_0^t \int_{-\infty}^{\infty} \frac{1}{\sqrt{4\pi(t-s)}} e^{-\frac{(x-y)^2}{4(t-s)}} (O(1)(s + 1)^{-3/2} e^{-\frac{(y-(s+1))^2}{4(s+1)}})_y dy ds$$

$$= \int_0^t O(1)(t - s)^{-1/2}(s + 1)^{-1}(t + 1)^{-1/2} e^{-\frac{(x-(s+1))^2}{D(t+1)}} ds.$$

Here we have used integration by parts before completing square and integrate with respect to y. The constant D can be any constant greater than 4. For $x < \sqrt{t+1}$

$$iv_1 = \int_0^t O(1)(t-s)^{-1/2}(s+1)^{-1}(t+1)^{-1/2}e^{-\frac{x^2}{D(t+1)}}e^{-\frac{(s+1)^2}{D(t+1)}}ds$$

$$= O(1)(t+1)^{-1/2}e^{-\frac{x^2}{D(t+1)}}\left[\int_0^{\sqrt{t+1}}(t+1)^{-1/2}(s+1)^{-1}ds\right.$$

$$+ \int_{\sqrt{t+1}}^{t/2}(t+1)^{-1/2}(s+1)^{-1}e^{-\frac{(s+1)^2}{D(t+1)}}ds + \int_{t/2}^t (t-s)^{-1/2}(t+1)^{-1}e^{-\frac{(t+2)^2}{D(t+1)}}ds\right]$$

$$= O(1)(t+1)^{-1}\ln(t+1)e^{-\frac{x^2}{D(t+1)}}.$$

For $\sqrt{t+1} < x < t - \sqrt{t+1}$

$$iv_1 = O(1)\left[\int_0^{x/M}(t+1)^{-1}(s+1)^{-1}e^{-\frac{x^2}{D(t+1)}}ds\right.$$

$$+ \int_{x/M}^{t-(t-x)/M}(t-x)^{-1/2}x^{(-1}(t+1)^{-1/2}e^{-\frac{(x-(s+1))^2}{D(t+1)}}ds$$

$$+ \int_{t-(t-x)/M}^t (t-s)^{-1/2}(t+1)^{-3/2}e^{-\frac{(x-(t+1))^2}{D(t+1)}}ds\right]$$

$$= O(1)\left[(t+1)^{-1}(\ln x)e^{-\frac{x^2}{D(t+1)}} + (t-x)^{-1/2}x^{-1} + (t+1)^{-1}e^{-\frac{(x-(t+1))^2}{D(t+1)}}\right].$$

The constant M is chosen large so that D is close to 4. For the case $|x-t| < \sqrt{t+1}$

$$iv_1 = O(1)\left[\int_0^{t/2}(t+1)^{-1/2}(s+1)^{-1}(t+1)^{-1/2}e^{-C(t+1)}ds\right.$$

$$+ \int_{t/2}^{t-\sqrt{t+1}}(t+1)^{-1/4}(t+1)^{-1}(t+1)^{-1/2}e^{-\frac{(x-(s+1))^2}{D(t+1)}}ds$$

$$+ \int_{t-\sqrt{t+1}}^t (t-s)^{-1/2}(t+1)^{-1}(t+1)^{-1/2}ds\right]$$

$$= O(1)(t+1)^{-1},$$

where C is a positive constant. Finally, for the case $x > t + \sqrt{t+1}$

$$iv_1 = \int_0^t O(1)(t-s)^{-1/2}(s+1)^{-1}(t+1)^{-1/2}e^{-\frac{(x-(t+1))^2}{D(t+1)}}e^{-\frac{(t-s)^2}{D(t+1)}}ds$$

$$= O(1)(t+1)^{-1/2}e^{-\frac{(x-(t+1))^2}{D(t+1)}}\left[\int_0^{t/2}(t+1)^{-1/2}(s+1)^{-1}e^{-\frac{(t-s)^2}{D(t+1)}}ds\right.$$

$$+ \int_{t/2}^{t-\sqrt{t+1}}(t-s)^{-1/2}(t+1)^{-1}e^{-\frac{(t-s)^2}{D(t+1)}}ds$$

$$+ \int_{t-\sqrt{t+1}}^t (t-s)^{-1/2}(t+1)^{-1}ds\right]$$

$$= O(1)(t + 1)^{-1}e^{-\frac{(x-(t+1))^2}{D(t+1)}}.$$

This completes the estimate of iv_1 and shows that it decays at the rates of (3.12).

The rational for the complex breakup of integral in estimating iv_1 can be understood in the following way. The diffusion waves propagate in the characteristic direction and diffuse with the width $\sqrt{s+1}$; the same with the Green function, which is ,though, in the negative time direction. This is the reason for the breakup of the integral with limits involving $\sqrt{t+1}$ in the above calculations. The estimates for the diffusion waves of algebraic types in v is more complicated, but follows from the same principle.

The non-resonance condition $j \neq i$ for the term ii is important in the accuracy of $\Theta(x,t)$ as time-asymptotic solution of (3.1). To make use of it we first notice that the fuctions $\theta_j{}^2(y,s)$ as defined in (3.5) dissipates along the characteristic direction λ_j and satisfies

$$(\theta_j{}^2)_s + \lambda_j(0)(\theta_j{}^2)_y = (\theta_j{}^2)_{yy} + O(1)\varepsilon^2(s+1)^{-2}e^{-\frac{(x-\lambda_j(s+1))^2}{D(s+1)}}, \qquad (3.14)$$

for any constant $D > 4$. The Green function

$$G_i(y,s) \equiv (4\pi(t-s))^{-1/2}e^{-\frac{(x-y-\lambda_i(t-s))^2}{4(t-s)}}$$

satisfies

$$G_{is} + \lambda_i G_{iy} + G_{iyy} = 0, \qquad G(x,t) = \delta(y-x). \qquad (3.15)$$

for any constant $D > 4$. With these estimates the term ii is dealt with the following way

$$ii = O(1)[\int_0^{\sqrt{t+1}} \int_{-\infty}^{\infty} G_{iy}(y,s)\theta_j{}^2(y,s)dyds$$

$$+ \int_{\sqrt{t+1}}^t \int_{-\infty}^{\infty} G(y,s)\frac{1}{\lambda_j - \lambda_i}[(\theta_j{}^2)_s + \lambda_j(\theta_j{}^2)_y - (\theta_j{}^2)_{yy}](y,s)dyds$$

$$- \int_{\sqrt{t+1}}^t \int_{-\infty}^{\infty} G_i(y,s)\frac{1}{\lambda_j - \lambda_i}[(\theta_j{}^2)_s + \lambda_i(\theta_j{}^2)^2)_y - (\theta_j{}^2)_{yy}](y,s)dyds]$$

$$\equiv ii_1 + ii_2 + ii_3.$$

The integrant in ii_1 is the same as before, but the integal limit is only up to $\sqrt{t+1}$, and so the integral decay faster than it would be if it is integrated up to the time t, e.g. the estimate for the term iv_1 above. Note, however, that, due to (3.14), the integrant in ii_2 has an additional decay rate of $(s+1)^{-1/2}$ as compared to that of the original form of ii. This again accounts for the sateisfactory decay rate. The term ii_3 is treated using (3.15) and integration by parts

$$ii_3 = -\frac{1}{\lambda_j - \lambda_i}\theta_j{}^2(x,t) + \int_{-\infty}^{\infty} G_i(y,\sqrt{t+1})\frac{1}{\lambda_j - \lambda_i}\theta_j{}^2(y,\sqrt{t+1})dy,$$

which is of the form (3.12). It follows from (3.12) easily that the term iii is $O(1)\varepsilon(t+1)^{-1/4}\theta_i(x,t)$, again of the form (3.12). For details see Section 4 of [Liu-Zeng].

3.3 Shock and Rarefaction Waves

Viscous shock waves are travelling waves of the viscous conservation laws. As with the scalar conservation law, such a wave results from the compression of characteristics and the smoothing effect of the viscosity. In the case of system there are more than one characteristic families. A travelling wave with speed s is called $p - shock$ if it results from the compression of p–characteristic field:

$$\lambda_p(u_-) \geq s \geq \lambda_p(u_+) \tag{3.16}_1.$$

Since the inviscid system is assumed to be strictly hyperbolic, (3.4), for weak shock, $|u_+ - u_-|$ small, the shock s speed also satisfies

$$s > \lambda_j(u_-), \quad s < \lambda_k(u_+), \quad j < p < k. \tag{3.16}_2$$

The above is the well-known Lax entropy condition, [Lax]. The shock wave (u_-, u_+) takes values along a curve, the Hugoniot curve, which is tangent to $r_p(u)$ at $u = u_-$. In particular, the vectors $u_+ - u_-, r_j(u_-), r_k(u_+), 1 \leq j < p < k \leq n$, form a basis in the state space R^n. Thus given any finite total mass (integral) perturbation of the shock

$$u(x,0) = \phi(x) + v(x,0), \quad \int_{-\infty}^{\infty} v(y,0)dy < \infty, \tag{3.17}$$

we have the unique decomposition

$$\int_{-\infty}^{\infty} v(y,0)dy \equiv \sum_{i<p} c_i r_i(u_-) + c_p(u_+ - u_-) + \sum_{i>p} c_i r_i(u_+). \tag{3.18}$$

This identity provides the basic information on the large-time behavior of the solution: A perturbation creats waves for each characteristic family. The p–waves are absorbed into the shock due to compression and gives rise to a phase shift of the shock, as in the scalar case. The phase shift creats a mass proportional to $u_+ - u_-$, also same as the scalar case. The i–waves, $i < p$, propagate toward the left and eventually diffuse around the base state u_- and create a mass proportional to $r_i(u_-)$ as we have seen in the last two sections. Similarly i–waves, $i > p$, create a mass proportional to $r_i(u_+)$. Thus the identity (3.18) yields the phase shift c_p of the shock, the amount of i–diffusion waves $c_i, i \neq p$. The stability analysis is much harder than that for the scalar case because of nonlinear interactions of waves of different characteristic families. Same with the diffusion waves case in the last two sections, if the shock is shifted by c_p and diffusion waves with strengh $c_i, i \neq p$, are substracted from the perturbation, then the remaining decays in $L_q(x), 1 \leq q \leq \infty$, as a result of pointwise decay. For details see [Liu3].

For the rarefaction waves we have seen in Section 1.3, there is no $L_1(x)$ decay for the perturbation even in the case of scalar equation. The analysis therefore is also quite different from the one just described for the shock waves. The rarefaction wave is stable because it is expansive. The amount of diffusion waves for the other families cannot be determined a priori as in (3.18). Another important difference is that the rarefaction wave cannot be defined explicitly as an exact solution of the system. Instead, it can only be defined time-asymptotically as follows: Suppose taht we have two states u_- and u_+ connected by the integral curve of the vector field $r_p(u)$. For the inviscid case, the rarefaction wave requires the characteristic speed $\lambda_p(u)$ to be strictly increasing as u moves along the integral curve from u_- to u_+, i.e., the p–field is $(g.nl)$. If we normalized the vector $r_p(u)$ so that $\nabla\lambda_p \cdot r_p(u) = 1$ then the time-asymptotic rarefaction wave for the viscous system can be defined by requiring it to lie on the integral curve and by prescribing $\lambda_p(u)$ to satisfy the Burgers equation

$$\lambda_t + \lambda\lambda_x = \lambda_{xx}, \quad \lambda(x,t) \equiv \lambda_p(u(x,t)).$$

The truncatin error of this approximation is

$$u_t + f(u)_x - u_{xx} = \lambda_t r_p(u) + f'(u)\lambda_x r_p(u) - \lambda_x r_p(u)_x$$

$$= (\lambda_t + \lambda\lambda_x - \lambda_{xx})r_p(u) - \lambda_x r_p(u)_x$$

$$= O(1)(\lambda_x)^2.$$

The error decays as time tends to infinity at sufficient fast rate to qualify the approximation as accurate time-asymptotic solution. However, this error also creats slower decay for the perturbation, [Szepessy-Zumbrun].

The stability analysis for the shock and rarefaction waves involves study of waves coupling, some of which have been described in Section 3.2, as well as the study of Green functions for the linearized system around the nonlinear waves. The latter is trivial for the study of diffusion waves in Sections 3.1 and 3.2, since the linearized system in that case is of constant coefficients and its Green function is simply the linear superposition of linear heat kernels. For the shock and rarefaction waves we need to study the Green functions for the linear equations with variable coefficients. The stability of viscous shocks is done by constructing approximate Green functions, accurate for small, intermediate and large times. Since the approximate rarefaction wave is based on the Burgers equation, the Green function for linearized system has explicit expression.

The interesting case of the stability of wave pattern consisting of both shock and rarefaction waves can now be studied using the pointwise approach. There is a major difficulty in that these two types of waves are stable in the distinct senses. The time-asymptotic error for the approximate rarefaction waves aforementioned produces strong nonlinear coupling between waves of different families. This results in the large deviation of the shock location form its coresponding inviscid ones, [Liu5]. The pointwise approach based on the Green functions for the linearized system yields several striking new phenomena and also raises interesting new problems for linear PDE.

4. System of Viscous Conservaton Laws, II

Consider the general system of viscous conservatin laws

$$u_t + f(u)_x = (B(u)u_x)_x. \tag{4.1}$$

There are two important simplifications made in the last section. The first is that the viscosity matrix $B(u)$ is tsken as an identity matrix. The second is that the corresponding inviscid system was assumed to be strictly hyperbolic. For physical systems the viscosity matrix is degenerate and not positive definite. This is so for the simplest case of the compressible Navier-Stokes equation. When this is the case, the solution operator is less disspative, or only partially dissipative. Many systems, including MHD and the full elasticty equations are not strictly hyperbolic. Such a system possesses new types of waves with richer qualitative behaviours. Though we may not require the inviscid system to be strictly hyperbolic, it is assumed to be completely hyperbolic, that is, the right eigenvectors $r_i(u)$, $i = 1, 2, \cdots, n$, form a basis of the state space R_n, even though the eigenvalues $\lambda_1(u) \leq \lambda_2(u) \leq \cdots \leq \lambda_n(u)$ may not be distinct, (1.3).

We first study the effect of the nonstrict hyperbolicity on the shock waves. In the next two subsections we consider the overcompressive and undercompressive shocks. For simplicity, the viscosity matrix is taken to be the identity matrix.

The two issues of nonstrict hyperbolicity and physical viscosity have been resolved satisfactorily for the diffusion waves. We describe this in the last subsection, which demands the basic dissipation requirements of the coupling of the flux $f(u)$ and the viscosity matrix $B(u)$.

4.1 Overcompressive Shocks

When the inviscid system is not strictly hyperbolic, there may exist shock waves which are more (or less) compressive than the calssical Lax shocks, (3.16). These are called the overcompressive (or undercompressive) shocks. These two types of waves have distinct behaviours. In this section we consider the overcompressive shocks, in the simplest setting of two conservation laws $n = 2$. Thus we have two characteristic values $\lambda_1(u) \leq \lambda_2(u)$. For the classical Lax 1−shock with speed s we have $\lambda_1(u_-) < s < \lambda_1(u_+) < \lambda_2(u_+)$, $s < \lambda_2(u_-)$; and for 2−shcok, $s > \lambda_1(u_+) > \lambda_2(u_+)$, $s < \lambda_2(u_-)$. In other words, of the four characteristics on either side of the shock, $\lambda_1(u_\pm), \lambda_2(u_\pm)$, there are three characteristics impinging on the shock and one leaving the shock. The shock is called overcompressive if it is more compressive than the classical Lax shock. In the case of two conservation laws this means that all characteristics impinge on the shock

$$\lambda_1(u_+) \leq \lambda_2(u_+) < s < \lambda_1(u_-) \leq \lambda_2(u_-). \tag{4.2}$$

As we shall see, the behavior of overcompressive shocks depends on the strengh of the viscosity. Consider the following two conservation laws

$$u_t + f(u)_x = \varepsilon u_{xx}. \tag{4.3}$$

The viscous $\phi((x - st)/\varepsilon)$ satisfies the ODE

$$-s(\phi(\xi) - u_\pm) + (f(\xi) - f(u_\pm)) = \phi'(\xi), \quad \phi(\pm\infty) = u_\pm,$$

$$u(x,t) = \phi(\xi), \quad \xi = \frac{x - st}{\varepsilon}. \tag{4.4}$$

For the calssical Lax 1–shock u_- is a nodal critical point and u_+ a saddle critical point of the ODE. For the classical Lax 2–shock the reverse holds. In either case, the connecting orbit is unique. The overcompressive shock, on the other hand, is a node-node connection since the sign of the eigenvalus of the left-hand side linearized around u_- (or u_+) is positive (or negative) by (4.2). Thus, when exists, there is a 1-parameter family of viscous profiles.

Since all characteristics impinge on the overcompressive shock, a perturbation cannot yields diffusion waves propagating away from the shock. The perturbation would have to propagate toward the shock and be cancelled, if the shock is indeed nonlinearly stable. On the other hand, since there is a 1-parameter family of viscous shocks with given end states $u_\pm$, the stability of a shock would have to be understood in the following way: The perturbation of a stable shock profile would converge to another profile in the 1-parameter family. Thus, in addition of the phase shift, the perturbation also change the time-asymptotic profile of the solution. Therefore, instead of using the conservation laws to identify the phase shift and diffusion waves as for the Lax shock, e.g. (3.17), we should use the two conservation laws to identify the phase shift and the new profile. This should make the situation well-posed as we have two conservatin laws and the same number of parameters to determine. However, we will see that this depends crucially on the strengh of the viscosity ε.

We illustrate this for the following simple rotationally invariant system originated from the study of MHD and nonlinear elasticity by Freistuler, see [Freistuhler-Liu] and refences therein,

$$u_t + (u(u^2 + v^2))_x = \varepsilon u_{xx},$$
$$v_t + (v(u^2 + v^2))_x = \varepsilon v_{xx}. \tag{4.5}$$

The characteristics are

$$r_1(u,v) = (v,-u), \quad r_2(u,v) = (u,v), \quad \lambda_1(u,v) = u^2 + v^2, \quad \lambda_2(u,v) = 3\lambda_1(u,v). \tag{4.6}$$

The notation $u^\perp$ denotes the vector orthogonal to u. The 1–characteristic is linearly degenerate and the 2-characteristic is genuinely nonlinear except at the origin

$$\nabla\lambda_1 \cdot r_1(u,v) = 0, \quad \nabla\lambda_2 \cdot r_2(u,v) = 6(u^2 + v^2). \tag{4.7}$$

A viscous shock wave has end states along the same radial direction through the origin, i.e. in the direction of $r_2(u,v)$. The system is rotational invariant and so, without loss of generality, consider the ends states to have the second component zero $(u_\pm, 0)$. When u_- and u_+ are of the same sign, $u_- > u_+ > 0$, the shock is a Lax shock and there is only one connecting orbit along the u-axis. When $u_+ < 0$, the shock may cross the point of non-strictly hyperbolic point $(u,v) = 0$ and becomes overcompressive. It is easy to see that connecting orbits exist if

$$0 > u_+ > u_0, \quad 3u_0^2 \equiv \frac{u_0^3 - u_-^3}{u_0 - u_-}. \tag{4.8}$$

When the end state $u_+ = u_*$ the shock fails to be overcompressive as in the case $0 < u_+ < u_-$. With (4.8) there is a 1-parameter family of orbits, Figure 1. These orbits are bounded by two the orbits connecting $(u_-, 0)$ and $(u_*, 0)$. with

$$\frac{u_0{}^3 - u_-{}^3}{u_0 - u_-} \equiv \frac{u_+{}^3 - u_-{}^3}{u_+ - u_-}, \quad u_* < u_0 < u_+.$$

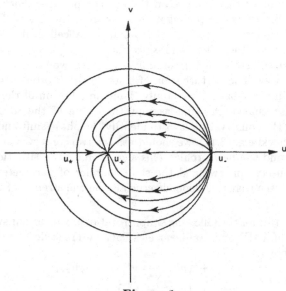

Figure 1.

Now consider the perturbation of one of the connecting orbits (ϕ_1, ψ_1)

$$(u, v)(x, 0) = (\phi, \psi)(\frac{x}{\varepsilon}) + (\overline{u}, \overline{v})(x, 0),$$

Figure 1. By earlier reasoning, stability means that there exists another profile (ϕ_2, ψ_2) such that time-asymptotically $(u, v)(x, t)$ tends to $(\phi_2, \psi_2)((x + x_0 - st)/\varepsilon)$ for some phase shift x_0. Moreover, since there is no diffusion waves, the convergence should be in any $L_p(x)$, $p \geq 1$ norm. This implies that their difference should converge to zero in these norms. Since we have conservation laws,

$$\int_{-\infty}^{\infty} ((u, v)(x, t) - (\phi_2, \psi_2)(\frac{x + x_0 - st}{\varepsilon}))dx = \int_{-\infty}^{\infty} ((\phi_1, \psi_1)(\frac{x - st}{\varepsilon})$$

$$+ (\overline{u}, \overline{v})(x, 0) - (\phi_2, \psi_2)(\frac{x + x_0 - st}{\varepsilon}))dx.$$

Thus the new profile (ϕ_2, ψ_2) and the phase shift x_0 are related to the perturbation $(\overline{u}, \overline{v})$ by the following

$$\int_{-\infty}^{\infty} (\overline{u}, \overline{v})(x, 0)dx = \int_{-\infty}^{\infty} ((\phi_2, \psi_2)(\frac{x + x_0 - st}{\varepsilon}) - ((\phi_1, \psi_1)(\frac{x - st}{\varepsilon}))dx. \quad (4.9)$$

Note that the phase shift x_0 contributes to the integral only for the first component, because the end states $(u, v)_\pm$ have zero second component. Consequently, the first integral in (4.9) can be satisfied with properly chosen x_0. Moreover, x_0 is chosen independent of ε. For simplicity, assume that $x_0 = 0$. We now study the feasibility of the second integral, we first rewrite the second equation in (4.9) as

$$\int_{-\infty}^{\infty} \bar{v}(x, 0)dx = \varepsilon \int_{-\infty}^{\infty} (\psi_2(\xi) - \psi_1(\xi)d\xi.$$

Since the trajectories connecting $(u_-, 0)$ and $(u_+, 0)$ form an open set in the state space, The second integral above can always be satisfied if the initial perturbation has sufficiently small total mass, i.e. the left-hand side of (4.9) is sufficiently small. However, unless the second integral on the right-hand side of (4.9) can be arbitraily large for varying (ϕ_2, ψ_2), the total mass of the second component of the perturbation has to be of the same size as the strength of the viscosity ε. The question is then whether the integral of the second component of the trajectories is uniformly bounded or not. It can be shown from (4.4) and (4.5) that the integral is indeed uniformly bounded by the finite integrals of the projactories connecting $(u_-, 0)$ and $(u_*, 0)$.

The above analysis shows that viscous profiles for the overcompressive shocks are nonlinearly stable, but not uniformly with respect to the strength of the viscosity. In other words, the overcompressive shocks for (4.5), the intermediate shocks for MHD, are physical provided that dissipations are not small. They are not admissible inviscid shocks. This is consistent with the study of the hyperbolic equations, as the inclusion of the intermediat shocks would make the Riemann problem not to have a unique solution. For details, see [Freistuhler-Liu] and refernces therein.

4.2 Undercompressive Shocks

An undercompressive shock has less characteristics impinge on the shock as compared to the Lax shock. Such a shock occurs in physical situations such as combustions. The weak detonations and weak deflagrations are undercompressive. They also occur in the multi-phase flows. For MHD and nonlinear elasticity, waves which share the similar properties as the undercompressive shocks also exist. One such example is the the two orbits connecting $(v_-, 0)$ and $(v_0, 0)$ in Figure 1 of the last example.

For two conservation laws, for instance, the number of characteristics on either side of the shock impinging on the shock is less than three. It turns out that the number can only be two in this case. When this happens, the viscous shock is a saddle-saddle connection of the ODE, cf. (4.4). As such, the right end state u_+ is uniquely determined by the left end state u_-, and vice versa. Moreover, this dependence is a function of the viscosity matrix $B(u)$. This is so, because saddle-saddle connection is not stable in the strict dynamic sense.

There is another basic difference between the undercompressive shocks and other two types of shocks considered previously. It is that the conservaton laws cannot be used to determine the phase shift of the shock due the perturbation. This is so because there are three time-asymptotic parameters, the phase shift of the shock and the strength of the two diffusion waves, corresponding to the two characteristics leaving the shock, to determine, and there are only two conservation laws at disposal. This makes it harder to come up with the stability analysis, if indeed it is stable.

For these reasons, it has been doubted that such a shock is nonlinearly stable. That they are stable has been shown for a class of systems. We outline the reasoning behind the analysis. The first seemingly obstacle aforementioned is offseted by the following two facts: First, although saddle-saddle connections are unstable, it is nevertheless generic, as one of the end states can always be found when the other is perturbed. Secondly, the number of characteristics leaving the shock is more than that for the Lax shocks. This makes it easier to find the time-asymptotic states. As to the second obstacle of not being able to find the phase shift using the conservation laws, this is now possible due to the pointwise approach introduced recently by the author, e.g. Sections 2 and 3 in this article. This makes it possible to study the time-asymptotic decoupling of waves of different families and allows for the time-asymptotic determination of the phase shift and therfore the stability analysis.

We now outline the approach for the two conservation laws

$$u_t + f(u)_x = (B(u)u_x)_x, \qquad (4.10)$$

with a stationary undercompressive shock

$$f(\phi(x)) - f(u_\pm) = B(\phi)\phi_x. \quad u_\pm \equiv \phi(\pm\infty). \qquad (4.11)_1$$

$$\lambda_1(u_\pm) < 0 < \lambda_2(u_\pm). \qquad (4.11)_2$$

Consider a perturbation of the shock

$$u(x,t) = \phi(x) + v(x,t). \qquad (4.12)$$

From above the perturbation satisfies

$$v_t = Lv + (O(1)[|v|^2 + |v_x|^2])_x, \qquad (4.13)$$

$$Lv \equiv ([- f'(\phi) + B'(\phi)\phi_x]v)_x + (B(\phi)v_x)_x. \qquad (4.14)$$

Consider the adjoint equation

$$L^*e \equiv e_x f'(\phi) + e_{xx} B(\phi) = 0. \qquad (4.15)$$

As mentioned earlier, the conservation laws cannot used to determine the phase shift in the case of undercompressive shocks. The idea is to locate the shock time-asymptotically. This is down in two steps: First, determine the shock location for the linearized equations with data at a given time t_0

$$w_t = Lw, \quad w(x,t_0) = w_0(x). \qquad (4.16)$$

Then we need to obtain the pointwise estimates, taking advantage of the accurate shock location based on the linearized equations, and to find out the rate of decoupling of the nonlinear terms. This would allow us to be able to trace the location of the shock as time evolves, again using the linear equations (4.16) with the initial data $w_0(x)$ to vary with the exact solution $v(x, t_0)$.

We now show how to locate the shock from (4.16). From (4.15) and (4.16) we have the conservation law

$$\frac{d}{dt} < e, w > = < e, Lw > = < L^*e, w > = 0. \tag{4.17}$$

We have used $<\,,\,>$ to denote the inner product of vectors. By counting the index of L^* at $x = \pm\infty$ one sees that (4.15) has nonconstant solutions only if the shock is undercompressive, $(4.11)_2$. Recall that the initial data $w_0(x)$ depends on the phase shift. The phase shift, together with the amount of diffusion waves propagating away from the shock carrying a mass proportional to $r_1(u_-)$ and $r_2(u_+)$, account for the total mass

$$\int_{-\infty}^{\infty} w_0(y)dy.$$

By combinig the nonconstant solution and constant solutions we can find a solution $e(x)$ of the adjoint equation (4.15) with the property

$$< e, r_1(u_-) > = < e, r_2(u_+) > = 0.$$

With this, the inner product $< e, w_0 >$ does not have the contribution of the outgoing diffusion waves, which we do not know a priori. Consequently, the shock location should be determined by

$$< e, w_0 > = 0.$$

This determination of shock location is exact for the linear equations for all time because of the conservation law (4.17). The complete analysis is contained in [Liu-Zumbrun].

4.3 Diffusion Waves for General System

Consider the general system of viscous conservation laws

$$u_t + f(u)_x = (B(u)u_x)_x, \tag{4.18}$$

where $u(x, t) \epsilon R^n$ and the characteristic values for the corresponding inviscid system are

$$\lambda_1(u) < \lambda_2(u) < \cdots, \lambda_s(u), \tag{4.19}$$

with multiplicity $m_1, m_2, \cdots, m_s$, $m_1 + m_2 + \cdot + m_s = n$. The corresponding eigenvectors are

$$f'(u)r_{ij}(u) = \lambda_i(u)r_{ij}(u), \quad l_{ij}(u)f'(u) = \lambda_i(u)l_{ij}(u),$$

$$l_{ij}r_{i'j'} = \delta_{ii'}\delta_{jj'} \quad i, i' = 1, \cdot, s, \quad j = 1, \cdot, m_i, \quad j' = 1, \cdot, m_{i'}. \tag{4.20}$$

The assumption for the dissipation of the system concerns the coupling of the viscosity matrix $B(u)$ and the flux matrix $f'(u)$, [Kawashima],

I. The system (4.18) has a strictly convex entropy $\eta(u)$. This means that there exists a pair of smooth functions $\eta(u)$ and $q(u)$, the entropy pairs, such that $\eta(u)$ is convex, $\eta'f' = q'$, and $\eta''B$ is symmetric and semi-positive definite.

II. There exists a smooth one-to-one mapping $u = f_0(\overline{u})$ such that the null space N of $\overline{B}(\overline{u}) \equiv B(f_0(\overline{u}))f_0'(\overline{u})$ is independent of $\overline{u}$. Moreover, $N^\perp$ is invariant under $f_0'(\overline{u})^t\eta''(f_0(\overline{u}))$, and $\overline{B}(\overline{u})$ maps R^n into $N^\perp$, where $N^\perp$ is the orthogonal complement of N.

III. The right eigenvectors of $f'(u)$ is not in the null space of $B(u)$.

The assumption **I** implies that system (4.18) is symmtrizable. This is necessary for the basic energy estimates and make the inviscid system completely hyperbolic, (4.19), (4.20). Consider the perturbaton of the zero state and decompose the solution of (4.18) in the right eigenvector directions,

$$u(x,t) \equiv \sum_{i=1}^{s}\sum_{j=1}^{m_i} u_{ij}(x,t)r_{ij}(0) = \sum_{i=1}^{s} u_i(x,t)r_i(0), \qquad (4.21)$$

$$r_i \equiv (r_{i1},\cdots,r_{im_i}).$$

Set

$$l_i \equiv \begin{pmatrix} l_{i1} \\ \cdot \\ \cdot \\ \cdot \\ l_{im_i} \end{pmatrix}, \quad l_i r_i = I_{m_i x m_i}, \quad l_i r_j = 0_{m_i x m_j}, \quad i \neq j. \qquad (4.22)$$

Linearize the equation (4.18) around the zero state and diagonalize the flux vector to obtain

$$u_{it} + \lambda_i(0)u_{ix} + \frac{1}{2}l_i(0)f''(0)(\sum_j r_j(0)u_j, \sum_j r_j(0)u_j)_x$$

$$= l_i(0)B(0)\sum_j r_j(0)u_{jxx} + (O(1)(|u||u_x| + |u|^3))_x, \quad i = 1,\cdots,s. \qquad (4.23)$$

Each of the above is a system of m_i equations corresponding to an eigenvalue $\lambda_i(0)$. The time-asymptotic state is defined by neglecting the coupling and higher order terms

$$\theta_{it} + \lambda_i(0)\theta_{ix} + \frac{1}{2}l_i(0)f''(0)(r_i(0)\theta_i, r_i(0)\theta_i)_x = l_i(0)B(0)r_i(0)\theta_{ixx}, \quad i = 1,\cdots,s, \qquad (4.24)$$

where θ_i is an m_i-vector. Note that (4.24) is a system of Burgers-like equations when its quadratic term is nonzero. It can be shown that the system has self-similar solutions, cf. (2.3), [Chern],

$$\theta_i(x,t) = t^{-1/2}\psi(xt^{-1/2}).$$

The time-asymptotic state is the combination of these solutions with the property that the solution minus it has zero total mass,cf. (2.9). The assumptionsI and II imply that (4.24) is uniformly parabolic. The above three assumptions allow for the energy method to be combined with the generalization of the pointwise estimates in Section 2.1. We then obtain similar decay estimates as (2.12). Basic to the approach is the pointwise estimates of the Green functions for the linearized system. The Green functions contain δ-functions since (4.18) is not uniformly parabolic. For details of the theory and its applications to physical systems see [Liu-Zeng].

5. Hyperbolic Conservation Laws with Relaxation

Relaxation is a phenonmenon which occurs in a wide varieties of physical situations. In gas dynamics, it occurs when the gas is in thermo-non-equilibrium. In elasticity it is usually reffered to as fading memory. Hyperbolic consevation laws with relaxation are also served as kinetic models. The following simple model of two equations captures the basic features of these physical models

$$u_t + f(u,v)_x = 0, \qquad (5.1)_1$$

$$v_t + g(u,v)_x = h(u,v). \qquad (5.1)_2$$

Here the first equation represents a conservation law. The second equation is the rate equation. The function $h(u,v)$ often takes the following form

$$h(u,v) = \frac{V(u) - v}{\varepsilon}. \qquad (5.2)$$

The function $V(u)$ is the equilibrium value for v and ε is the relaxation time, which is small in many physical situations. In the kinetic theory, the relaxation time is the mean free path, and in elasticity the length of the memory. More generally, (5.2) is replaced by

$$h_v(u,v) < 0, \quad h(u,V(u)) = 0, \qquad (5.2)'$$

for some function $V(u)$. The two equations in (5.1) are assumed to be coupled, i.e. $f_v(u,v) \neq 0$. For definiteness, we assume

$$f_v(u,v) < 0. \qquad (5.3)$$

Some of the basic issues for such a model are the stability, nonlinear waves, zero relaxation, $\varepsilon \to 0$, and other limits.

5.1 Subcharcteristic Condition

The system (5.1) is assumed to be strictly hyperbolic with characteristics λ_1 and λ_2

$$(\lambda - f_u)(\lambda - g_v) - f_v g_u = 0,$$

$$\lambda = \lambda_1(u,v), \lambda_2(u,v) \quad \lambda_1 < \lambda_2. \qquad (5.4)$$

As the relaxation time ε tends to zero, the solutions are supposed to converge to those of the equilibrium equation

$$u_t + F(u)_x = 0, \quad F(u) \equiv f(u, V(u)). \qquad (5.5)$$

When (5.1) is viewed as a kinetic model then the equation (5.5) corresponds to the compressible Euler equaitions in the kinetic theory. The charateristic value for the equilibrium equation is

$$\Lambda(u) \equiv F'(u) = f_u(u, V(u)) + f_v(u, V(u))V'(u). \tag{5.6}$$

A basic stability criterion is to require that the limiting equation should have slower speed of propagation Λ than those of the original system λ_1 and λ_2:

$$\lambda_1(u, V(u)) < \Lambda(u) < \lambda_2(u, V(u)). \tag{5.7}$$

This is the basic subcharateristic condition for stability. We will see that it is consistent with other stability criteria. The criterion is well-known for the linearized system, which turns out to be the telegrapher's equation.

5.2 Chapman-Enskog Expansion

There is the important Chapman-Enskog expansion in the kinetic theory in deriving the compressible Navier-Stokes equations in the hydrodynamics limit. The expansion, unlike the Hilbert expansion, is to expand the equations and not the solutions. We apply the same approach to the simple model (5.1). Instead of assuming the solution is in local equilibrium, $v = V(u)$, as was done in deriving the Euler equation (5.5), we set

$$v = V(u) + v_1, \tag{5.8}_1$$

with the higher-order correction v_1 determined by $(5.1)_2$ and (5.2)

$$v_1 = -\varepsilon(V(u)_t + g(u, V(u))_x).$$

The Euler equation (5.5) is now used, not as first term in the expansion, but only to set the characteristic direction

$$\partial_t + \Lambda\partial_x = 0.$$

These two identities yield

$$v_1 = -\varepsilon(-\Lambda(u)V'(u) + g_u(u, V(u)) + g_v(u, V(u))V'(u))u_x. \tag{5.8}_2$$

Plug (5.8) into $(5.1)_1$ we obtain the analogue of the compressible Navier-Stokes equations in the kinetic theory

$$u_t + F(u)_x = (\varepsilon(\lambda_2 - \Lambda)(\Lambda - \lambda_1)u_x)_x, \tag{5.9}$$

where we have used (5.4), (5.6) and some simplification through the Taylor expansion.

Notice that the subcharacteristic condition (5.7) makes (5.9) well-posed with positive viscosity. In the derivation of the Navier-stokes equations it is assumed that higher derivatives is smaller. Thus (5.9) is valid for diffusion waves, time-asymptotically and not in the zero relaxation limit as ε tending to zero. Since the relaxation induces disspation, as evidenced from the above Chapman-Enskog expansion, the techniques for the study of diffusion waves in Section 3.1 can be applied here to verify the time-asymptotic validity of (5.9) for (5.1). In fact, this thinking is applicable to a wide variety of physical models, including the Broadwell or other kinetic models, viscoelasticity with fading memory, etc. For the detailed analysis of the simple model (5.1) described above see [Liu1].

One can derive a viscous equation which is actually the zero relaxation limit of (5.1). This is the weakly nonlinear limit, an anlogue of the incompressible Navier-Stokes equations in the kinetic theory. The incompressible Navier-Stokes equations capture the large-time behaviour of weak compressible flows. Thus we consider the small perturbation of an equilibrium state for (5.1)

$$(u^\varepsilon, v^\varepsilon) = (u_0, V(u_0)) + \varepsilon(w^\varepsilon, z^\varepsilon) + O(\varepsilon^2). \tag{5.10}$$

Here we have used the supersript ε to indicate the dependence of the solutions on the relaxation time. We next choose a moving frame with the equilibrium speed and a long-time scaling

$$(x, t) \longmapsto (x - \Lambda(u_0)t, \varepsilon t), \tag{5.11}$$

so that the base equilibrium speed becomes zero and the subcharacteristic condition becomes

$$\Lambda(u_0) = 0 > \lambda_1 \lambda_2 \tag{5.12}.$$

In consistent with most physical situations, we assume that the Euler equation (5.5) is nonlinear

$$\Lambda'(u) \neq 0. \tag{5.13}$$

Plug (5.10) into (5.1) with the new scaling (5.11) and use (5.12) and (5.13) we have the following equation as the leading-order term in the expantion with small parameter ε

$$-h_v(u_0, V(u_0))(w_t + \Lambda'(u_0)(\frac{w^2}{2})_x) = -\lambda_1(u_0, V(u_0))\lambda_2(u_0.V(u_0))w_{xx}. \tag{5.14}$$

As with the incompressible Navier-Stokes equations, the nonlinearity in (5.14) in quadratic, as the result of the expansion of the convex function $F(u)$, (5.13), around an equilibrium state $(u_0, V(u_0))$.

It can be shown again using the parabolic method of Sections 3 and 4 that (5.14) is the zero relaxation limit of (5.10) in the above setting.

5.3 Entropy

Mathematically the entropy pair$(\eta, q)(u, v)$ for two conservatin laws of the left-hand side of (5.1) satisfies

$$(\eta_u, \eta_v) \begin{pmatrix} f_u & f_v \\ g_u & g_v \end{pmatrix} = (q_u, q_v), \tag{5.15}$$

so that for smooth solutions of (5,1) we have, by multiplying $(5.1)_1$ by η_u and $(5.1)_2$ by η_v,

$$\eta_t + q_x = \eta_v h(u, v) \tag{5.16}_1,$$

[Lax 2]. For hyperbolic conservation laws the above is an inequality when there are shocks in the solution

$$\eta_t + q_x \leq \eta_v h(u, v) \qquad (5.16)_2.$$

Since the relaxation has the effect of smoothing, as indicated by the Chapman-Enskog expansion, the dissipation has to come from the negativeness of the right-hand side of (5.15)

$$\eta_v(u, v) h(u, v) < 0, \quad \text{unless } (u, v) \text{ is on the equilibria } h(u, v) = 0. \qquad (5.17)$$

We will see that this is so if and only if the subcharateristic condition (5.7) holds. We usually require the entropy function to be convex so that (5.16) provides useful energy estimate

$$\eta''(u, v) > 0. \qquad (5.18)$$

The identity (5.15) is a compatibility condition. It requires the entropy function $\eta(u, v)$ to satisfy

$$g_u \eta_{vv} - (g_v - f_u)\eta_{uv} - f_v \eta_{uu} = 0. \qquad (5.19)$$

This is a linear hyperbolic PDE. The equilibria $h(u, v) = 0$ is non-characteristic if

$$g_u + (g_v - f_u)V' - f_v V'^2 \neq 0.$$

It follows immediately from (5.3), (5.4) and (5.6) that this is implied by the subcharacteristic condition (5.7)

$$(\Lambda - \lambda_1)(\Lambda - \lambda_2) = f_v[f_v V'^2 + (f_u - g_v)f_v V' - g_u]. \qquad (5.20)$$

Thus we may impose initial conditions for $\eta(u, v)$ along the equilibria

$$\eta(u, V(u)) = \eta_0(u), \quad \eta_v(u, V(u)) = 0. \qquad (5.21)$$

For the first condition to be consistent with (5.18) we require

$$\eta_0''(u) > 0. \qquad (5.18)'.$$

The second condition in (5.21) is consistent with (5.17). Since the equilibrium set is non-characteristic for (5.19), it can be solved with the initial data (5.21). We now show that (5.18)' and the subcharateristic condition (5.7) implies the convexity of η (5.18) and the stability condition (5.16). Differentiating the first condition of (5.21) twice and the second one once with respect to u to obtain

$$\eta_{vu} + \eta_{vv}V' = 0, \qquad (5.22)_1$$

$$\eta_{uu} + 2\eta_{uv}V' + \eta_{vv}V'^2 = \eta_0'' > 0. \qquad (5.22)_2$$

From (5.19) and (5.22) we have

$$\eta_{uu} = \eta_{vv}V'^2 + \eta_0'',$$

$$\eta_{vv}[-V'^2 - \frac{f_u - g_v}{f_v}V' + \frac{g_u}{f_v}] = \eta_0'' > 0.$$

This and (5.20) imply that the convexity of the entropy $\eta(u,v)$ (5.18) is equivalent to the subcharacteristic condition (5.7). The stability condition now follows from (5.7) due to $\eta_{vv} > 0$, as a result of the convexity of η, and the relaxation condition $h_v < 0$, (5.2)'.

The construction of dissipative entropy pairs allows us to obtain apriori estimates. This enable us to apply the theory of compensated compactness to study the zero relaxation limit of the convergence of solutions of (5.1) to that of the Euler equation (5.5). For this and other issues in Section 5.2 see [Chen-Levermore-Liu].

5.4 Nonlinear Waves

The relaxation induces disspation, which, however, may not be sufficient to smooth out all the discontinuities in the solution, even time-asymptotically. When a relaxation model is semilinear, as in the case of Broadwell in the kinetic theory, smooth initial data give rise to smooth solutions. Nonsmoothness in the intial data persists to later time. This is so also for the linear systems. When a relaxation model is quasilinear, as in the case of gas dynamics equations in thermo-non-equilibrium, smooth intial data may give rise to shocks at later time. When the initial data and the derivatives are small, smooth global solutions persist due to the dissipation by relaxation. Small initial data with large derivatives give rise to shocks at later time, but these shocks will decay, again as the result of dissipation. Thus relaxation has complete disspation, time-asymptotically, for solutions with small oscillation. Consequently, that the semilinearity or quasilinearity of the model is not so important except for short-time behaviour of the solutions. There is another, far more important, consideration in the behaviour of the solutions. This concerns certain resonance phenomenon, which set apart models such as the Broadwell model and the gas dynamics equations in thermo-non-equilibrium. To understand it, we consider the nonlinear waves for (5.1).

Frist we study the travelling waves $(u,v)(x,t) - (\phi,\psi)(x-st)$. From (5.1), we have

$$-s(\phi - u_\pm) + f(\phi,\psi) - f(u_\pm,v_\pm) = 0, \tag{5.23}$$

$$-s\psi' + g(\phi,\psi)' = h(\phi,\psi). \tag{5.24}$$

Since the end states $u_\pm$ should be constant solutions of (5.1) and the only constant solutions are equilibrium states, we have from (5.23) the following jump condition

$$s = \frac{F(u_+) - F(u_-)}{u_+ - u_-}, \quad v_\pm = V(u_\pm). \tag{5.25}$$

From (5.23), (5.24), after straightforward calculations, we have

$$\phi' = \frac{-f_v h}{(\lambda_1 - s)(\lambda_2 - s)}, \tag{5.26}$$

$$\psi' = \frac{(f_u - s)h}{(\lambda_1 - s)(\lambda_2 - s)}. \tag{5.27}$$

When the shock is weak we have from the jump condition (5.25) that the shock speed s is close to the equilibrium speed Λ. By the subcharacteristic condition (5.7) we see that (5.26) and (5.27) have only smooth solutions, if they exist. The existence of the wave with given end states $(u_\pm, V(u_\pm))$ is guaranteed if the following Oleinik's entropy condition for the scalar equation (5.5) is satisfied

$$s < \frac{F(u) - F(u_-)}{u - u_-}, \quad \forall u \text{ between } u_- \text{ and } u_+. \tag{5.28}$$

In other words, the Euler equation (5.5) completely characterizes the far fields of the travelling waves of the full system (5.1). For a strong travelling wave, its speed may be far from the equilibrium characteristic Λ and equals the frozen speed λ_1 or λ_2. When this happens it follows from (5.26) and (5.27) that no smooth travelling wave exists. When this happens, the travelling wave contains shocks for the system (5.1). The shocks for (5.1) are the same as those for the corresponding homogeneous hyperbolic conservation laws

$$u_t + f(u, v)_x = 0,$$

$$v_t + g(u, v)_x = 0.$$

The existence of shocks for the hyperbolic conservation laws requires the generalization of the entropy condition (5.28). It turns out that, in spite of this complication, a travelling wave for (5.1) exists still under the entropy condition (5.28).

That strong travelling waves contain shocks may or may not happen, depending on specific physical models. For the Broadwell model in the kinetic theory it does not occur. For gas dynamics in thermo-non-equilibrium and nonlinear elasticity with fading memory, this occurs, see [Liu1] and references therein. With the travelling waves in place of shocks for the Euler equation (5.7), we can now construct the time-asymptotic states for solutions of (5.1) with given end states u_l at $x = -\infty$ and u_r at $x = \infty$. First note that a rarefaction wave $u(x, t)$ for (5.7) is expansive and therefore its equilibria $(u, V(u)(x, t)$ may be viewed as time-asymptotic solutions for (5.1). Thus we may first solve the Riemann problem (5.7) and

$$u(x, 0) = \begin{cases} u_l, & x < 0, \\ u_r, & x > 0, \end{cases}$$

The aymptotic state for (5.1) is obtained from this Riemann solution by substituting shocks by travelling waves and equilibrium rarefcation waves. The above analysis on travelling waves shows that when the end states u_l and u_r are not close then the solution may develope non-decaying shocks. In other words, the relaxation may not smooth the solutions, even time-asymptotically.

6. Source and Damping

An important physical situation involving the source is the combustions. The chemical reactions create rich and complex wave phenomena. There are weak and strong detonation and deflagration waves in combutions. Strong detonation waves

have been studied by Zeldovich, von Neumann and Doring, and similar to the
nonlinearly stable gas dynamics shocks. Strong deflagration waves are not physical
and not observed. The question of the stability of weak detonation and deflagration
waves remains open, though believed to be stable. We look at this problem from
our point of view in the preceeding sections. Consider the followig simple model,
[Fricket and Davis]

$$(qz + u)_t + f(u)_x = \varepsilon u_{xx}, \tag{6.1}_1$$

$$z_t = -K\phi(u)z. \tag{6.1}_2$$

Here the variable u represents the lumped gas variables and z the reactant. The
function ϕ satisfies

$$\phi(u) = \begin{cases} 0, & \text{for } u < 0, \\ > 0, & \text{for } 1 > u > 0, , \\ 1 & \text{for } u > 1 \end{cases}$$

$$\phi'(u) > 0, \quad \text{for } 1 > u > 0. \tag{6.2}$$

Thus we have set the ignition temperature to be $u = 0$. The value $z = 0$ is the
burnt state and $z = 1$ the unburnt state. The heat release q, the viscosity ε and the
reaction rate K are positive constants. K is usually large and ε small. Set

$$K\varepsilon \equiv \beta. \tag{6.3}$$

We will only study the detonation waves. This corresponds to the following assumption for (6.1)

$$f'(u) > 0, \quad f''(u) > 0. \tag{6.4}$$

The detonation waves are combustion waves, the travelling waves with burnt state
at the left and unburnt state at the right

$$(u, z)(x, t) = (U, Z)(x - st), \quad U(-\infty) < 0, U((\infty) > 0, Z(-\infty) = 0, Z(\infty) = 1. \tag{6.5}$$

Plug this into $(6.1)_1$ and integrate, we obtain the jump condition relating the wave
speed s to the end states

$$s = \frac{f(u_+) - f(u_-)}{u_+ + q - u_-}. \tag{6.6}$$

It is easy to see that in order to have a solution for (6.5) and (6.6) the wave must
be compressive at the right state

$$s > f'(u_+). \tag{6.7}$$

At the left state u_- there are three possibilities.

Case 1. $s > f'(u_-)$.

Case 2. $s = f'(u_-)$.

Case 3. $s < f'(u_-)$.

These cases correspond, respectively, to strong, Chapmann-Jouget and weak
detonation waves. The strong detonation is not monotone and can be viewed as a gas
dynamics shock followed by reaction wave, the ZND waves. They are as compressive
as the gas dynamics shocks. The weak detonation waves are undercompressive and

are saddle-saddle connections for the ODEs for the travelling waves of (6.1). By considering the time-asymptotic states as was done in section 5.4, it can be seen that there exists such a state consisting of these detonation waves. Thus these waves are generic and should be stable. That which kind of detonation waves is in the asymptotic state with given end states at $x = \pm\infty$ depends on the viscosity parameter, in this case β of (6.3). This is inconsistent with the general observation of undercompressive shocks in Section 4.2 that the inviscid theory depends on the viscosity matrix.

Finnaly, we remark that resonance also plays an important role in situations where source or damping is present. In the case of combustions, we have, for the Chapmann-Jouget detonations, the wave speed equals the fluid speed and this accounts for the interesting wave behavoiurs as noted above. We give another example involving the damping. Consider the following model for gas flow through the porous media

$$\rho_t + (\rho u)_x = 0,$$

$$(\rho u)_t + (\rho u^2 + p)_x = -\alpha\rho u. \tag{6.8}$$

The quantities ρ, u, $p = \rho^\gamma$ are, respectively the density, velocity, and the pressure of the gas. α is the friction constant and is positive. The interesting case is with the vacuum $\rho = 0$, which give rise to the resonance as the sound speed is zero there and the two charateristics for the system are equal. The interest is to study the coupling of this with the effect of damping. As it turns out the damping generates a quite different kind of dissipation from that of the viscosity and relaxation. Suppose that the gas has finite mass, ρ has compact support. Then the solution would approach those of the nonlinear diffusion equation

$$\rho_t = \frac{1}{\alpha}(\rho^\gamma)_{xx}, \tag{6.9}$$

when the second equation in (6.8) is replaced with the Darcy's law

$$p_x = -\alpha\rho u.$$

That this would be true is supported by the following family of particular solutions for (6.8) of the form

$$c^2(x, t) = e(t) - b(t)x^2, \quad u(x, t) = a(t)x,$$

where $c = \sqrt{\gamma\rho^{\gamma-1}}$ is the sound speed. These are solutions of (6.8) and approach Barenbaltt's self-similar solutions of the porous media equation (6.9), time-asymptotically, if

$$e' + (\gamma - 1)ea = 0,$$

$$b' + (\gamma + 1)ab = 0,$$

$$a' + a^2 + \alpha a - \frac{2}{\gamma - 1}b = 0.$$

cf. [Liu2].

We conclude by remarking that the interaction of damping, reaction and viscosity can give rise to rich wave phenomena, particularly when certain resonance is present. The present understanding of the subject is still elementary. There are the problems of zero dissipation, i.e. to understand the behaviour of solutions to viscous conservation laws as the strength of viscosity becomes small, and the study of systems which possess only partial disspation mechanisms. The latter occurs, for instance, in the compressible Navier-Stokes equations when the heat conductivity is small compared to the viscosity, and also in the compressible Euler equations with thermo-non-equilibrium.

References.

[Chen-Levermore-Liu], Chen, G.-Q., Levermore, D., and Liu, T.-P., Hyperbolic conservation laws with stiff relaxation terms and entropy. Comm. Pure Appl. Math. (1994).

[Chern], Chern, I.-L., Multiple-mode diffusion waves for viscous nonstrictly hyperbolic conservation laws, Comm. Math. Phys., 138 (1991) 51-61.

[Freistuhler-Liu], Freistuhler, H., and Liu, T.-P., Nonlinear stability of overcompressive shock waves in a rotational invariant system of viscous conservation laws, Comm. Math. Phys., 153 (1993) 147-158.

[Fricket-Davis], Fricket, W. and Davis, W.C., Detonations, University of California Press, Berkeley, 1979.

[Kawashima], Kawashima, K., Large-time behavior of solutions to hyperbolic-parabolic systems of conservation laws and applications, Proc. Roy. Soc. Edinburgh 106A(1987) 169-194.

[Lax], Lax, P.D., Hyperbolic System of Conservation Laws and the Mathematical Theory of Shock Waves, SIAM, Philadelphia, 1973.

[Liu 1], Liu, T.-P., Hyperbolic conservation laws with relaxation, Comm. Math. Phys. 108 (1987) 153-175.

[Liu 2], Liu, T.-P., Nonlinear hyperbolic-parabolic partial differential equations, Nonlinear Analysis, Proceedings, 1989 Conference. Liu, F.-C., and Liu, T.-P., (eds.), pp. 161-170. Academia Sinica, Taipei, R.O.C.; World Scientific.

[Liu 3], Liu, T.-P., Pointwise convergence to shock waves for the system of viscous conservation laws, (to appear.)

[Liu 4], Liu, T.-P., Interaction of nonlinear hyperbolic waves, Nonlinear Analysis, Proceedings, 1989 Conference. Liu, F.-C., and Liu, T.-P., (eds), pp. 171-184. Academia Sinica, Taipei, R.O.C.; World Scientific.

[Liu 5], Liu, T.-P., The Riemann problem for viscous conservation laws, (to appear).

[Liu-Zeng], Liu, T.-P., Large time behaviour of solutions for general quasilinear hyperbolic-parabolic systems of conservation laws, (preprint.)

[Liu-Zumbrun], Liu, T.-P., and Zumbrun, K., On nonlinear stability of general undercompressive viscous shock waves, (preprint.)

[Szepessy-Zumbrun], A. Szepessy, and Zumbrun, K., Stability of viscous conservation laws, Archive Rational Mech. Anal. (to appear.)

[Whitham], Whitham, J. Linear and Nonlinear Waves, New York: Wiley 1974.

CIME Session on "Recent Mathematical Methods in Nonlinear Wave Propagation"

List of Participants

D. AMADORI, SISSA, Via Beirut 2-4, 34014 Trieste, Italy
A. AROSIO, Dip.to di Matematica, Via M. D'Azeglio 85, 43100 Parma, Italy
S. BELLOMO, Dip.to di Matematica, Via Archirafi 34, 90123 Palermo, Italy
E. CALLEGARI, Scuola Normale Superiore, Piazza dei Cavalieri 7, 56126 Pisa, Italy
C. CATTANI, Dip.to di Matematica, Univ. La Sapienza, P.le A. Moro 2, 00185 Roma, Italy
V. CIANCIO, Dip.to di Matematica, Univ. di Messina, Contrada Papardo, Salita Sperone 31,
          98166 Sant'Agata, Italy
V. A. CIMMELLI, Dip.to di Matematica, Univ. della Basilicata, Via N. Sauro 85, 85100 Potenza, Italy
S. CLAUDI, Dip.to di Matematica, Univ. La Sapienza, P.le A. Moro 2, 00185 Roma, Italy
S. COLLINS, Dept. of Math. Physics, Univ. College Dublin, Belfield, Dublin 4, Ireland
C. CRASTA, SISSA, Via Beirut 2-4, 34014 Trieste, Italy
M. DOLFIN, Dip.to di Matematica, Univ. di Messina, Contrada Papardo, Salita Sperone 31,
          98166 Sant'Agata, Italy
A. DONATO, Dip.to di Matematica, Univ. di Messina, Contrada Papardo, Salita Sperone 31,
          98166 Sant'Agata, Italy
G. FERRARESE, Dip.to di Matematica, Univ. La Sapienza, P.le A. Moro 2, 00185 Roma, Italy
D. FOSCHI, Dip.to di Matematica, Piazza di Porta S. Donato 5, 40127 Bologna, Italy
F. FRANCHI, Dip.to di Matematica, Piazza di Porta S. Donato 5, 40127 Bologna, Italy
H. FREISTUHLER, Inst. f. Math. RWTH Aachen, Templertgraben 55, D-52056 Aachen, Germany
G. GEMELLI, Via al Quarto Miglio 24, 00100 Roma, Italy
M. GHISI, Scuola Normale Superiore, Piazza dei Cavalieri 7, 56126 Pisa, Italy
A. GRECO, Dip.to di Matematica, Via Archirafi 34, 90128 Palermo, Italy
M. HANLER, Univ. Rostock, FB Mathematik, Universitatsplatz 1, D-18051 Rostock, Germany
A. LYASHENKO, Istituto di Analisi Globale, CNR, Via S. Marta 13/A, 50139 Firenze, Italy
S. MEYER, Aufenauer Str. 36, D-63607 Wachtersbach, Germany
A. MORRO, DIBE, Università, Via Opera Pia 11/A, 16145 Genova, Italy
M.K.V. MURTHY, Dip.to di Matematica, Via Buonarroti 2, 56127 Pisa, Italy
M.C. NUCCI, Dip.to di Matematica, Via Vanvitelli 1, 06123 Perugia, Italy
A. PALUMBO, Dip.to di Matematica, Univ. di Messina, Contrada Papardo, Salita Sperone 31,
          98166 Sant'Agata, Italy
S. PANIZZI, Dip.to di Matematica, Via D'Azeglio 85/A, 43100 Parma, Italy
B. PICCOLI, SISSA, Via Beirut 2-4, 34014 Trieste, Italy
I. QUANDT, Inst. f. Reine Math., Humboldt Univ., Ziegelstr. 13a, D-10099 Berlin, Germany
V. ROMANO, Dip;.o di Matematica, Viale A. Doria 6, 95125 Catania, Italy
A. ROSSANI, Dip.to di Matematica del Politecnico, C.so Duca degli Abruzzi 24, 10129 Torino, Italy
B. RUBINO, Dip.to di Matematica, Via E. Orabona 4, 70125 Bari, Italy
R. SAMPALMIERI, Dip.to di Energetica, Località Monteluco, 67040 Roio Poggio (AQ), Italy
K. SAMUELSSON, Num. Anal. and Comp. Science, Royal Inst. of Technology,
          Lindestedsvagen 15-25, S-100 44 Stockholm, Sweden
P. SECCHI, Dip.to di Matematica, Via F. Buonarroti 2, 56127 Pisa, Italy
L. SECCIA, CIRAM, Via Saragozza 8, 40123 Bologna, Italy
L. SETA, Dip.to di Matematica, Via Archirafi 34, 90123 Palermo, Italy
L. STAZI, Dip.to di Matematica, Univ. La Sapienza, P.le A. Moro 2, 00185 Roma, Italy
A. TERRACINA, Dip.to di Matematica, Univ. La Sapienza, P.le A. Moro 2, 00185 Roma, Italy
G. VALENTI, Dip.to di Matematica, Univ. di Messina, Contrada Papardo, Salita Sperone 31,
          98166 Sant'Agata, Italy

1972 – 59. Non-linear mechanics   "
       60. Finite geometric structures and their applications   "
       61. Geometric measure theory and minimal surfaces   "

1973 – 62. Complex analysis   "
       63. New variational techniques in mathematical physics   "
       64. Spectral analysis   "

1974 – 65. Stability problems   "
       66. Singularities of analytic spaces   "
       67. Eigenvalues of non linear problems   "

1975 – 68. Theoretical computer sciences   "
       69. Model theory and applications   "
       70. Differential operators and manifolds   "

1976 – 71. Statistical Mechanics   Ed Liguori, Napoli
       72. Hyperbolicity   "
       73. Differential topology   "

1977 – 74. Materials with memory   "
       75. Pseudodifferential operators with applications   "
       76. Algebraic surfaces   "

1978 – 77. Stochastic differential equations   "
       78. Dynamical systems   Ed Liguori, Napoli and Birhäuser Verlag

1979 – 79. Recursion theory and computational complexity   "
       80. Mathematics of biology   "

1980 – 81. Wave propagation   "
       82. Harmonic analysis and group representations   "
       83. Matroid theory and its applications   "

1981 – 84. Kinetic Theories and the Boltzmann Equation   (LNM 1048) Springer-Verlag
       85. Algebraic Threefolds   (LNM 947)   "
       86. Nonlinear Filtering and Stochastic Control   (LNM 972)   "

1982 – 87. Invariant Theory   (LNM 996)   "
       88. Thermodynamics and Constitutive Equations   (LN Physics 228)   "
       89. Fluid Dynamics   (LNM 1047)   "

```
1983 - 90. Complete Intersections              (LNM 1092) Springer-Verlag
       91. Bifurcation Theory and Applications (LNM 1057)         "
       92. Numerical Methods in Fluid Dynamics (LNM 1127)         "

1984 - 93. Harmonic Mappings and Minimal Immersions (LNM 1161)   "
       94. Schrödinger Operators               (LNM 1159)         "
       95. Buildings and the Geometry of Diagrams (LNM 1181)      "

1985 - 96. Probability and Analysis            (LNM 1206)         "
       97. Some Problems in Nonlinear Diffusion (LNM 1224)        "
       98. Theory of Moduli                    (LNM 1337)         "

1986 -  99. Inverse Problems                   (LNM 1225)         "
       100. Mathematical Economics             (LNM 1330)         "
       101. Combinatorial Optimization         (LNM 1403)         "

1987 - 102. Relativistic Fluid Dynamics        (LNM 1385)         "
       103. Topics in Calculus of Variations   (LNM 1365)         "

1988 - 104. Logic and Computer Science         (LNM 1429)         "
       105. Global Geometry and Mathematical Physics (LNM 1451)   "

1989 - 106. Methods of nonconvex analysis      (LNM 1446)         "
       107. Microlocal Analysis and Applications (LNM 1495)       "

1990 - 108. Geoemtric Topology: Recent Developments (LNM 1504)    "
       109. H  Control Theory                  (LNM 1496)         "
            ∞
       110. Mathematical Modelling of Industrical (LNM 1521)      "
            Processes

1991 - 111. Topological Methods for Ordinary   (LNM 1537)         "
            Differential Equations
       112. Arithmetic Algebraic Geometry      (LNM 1553)         "
       113. Transition to Chaos in Classical and (LNM 1589)       "
            Quantum Mechanics

1992 - 114. Dirichlet Forms                    (LNM 1563)         "
       115. D-Modules, Representation Theory,   (LNM 1565)         "
            and Quantum Groups
       116. Nonequilibrium Problems in Many-Particle (LNM 1551)   "
            Systems
```

1993 – 117. Integrable Systems and Quantum Groups   (LNM 1620)   Springer-Verlag
        118. Algebraic Cycles and Hodge Theory   (LNM 1594)
        119. Phase Transitions and Hysteresis   (LNM 1584)   "

1994 – 120. Recent Mathematical Methods in   (LNM 1640)   "
            Nonlinear Wave Propagation
        121. Dynamical Systems   (LNM 1609)   "
        122. Transcendental Methods in Algebraic   to appear   "
            Geometry

1995 – 123. Probabilistic Models for Nonlinear PDE's   (LNM 1627)   "
        124. Viscosity Solutions and Applications   to appear   "
        125. Vector Bundles on Curves. New Directions   to appear   "

Vol. 1594: M. Green, J. Murre, C. Voisin, Algebraic Cycles and Hodge Theory. Torino, 1993. Editors: A. Albano, F. Bardelli. VII, 275 pages. 1994.

Vol. 1595: R.D.M. Accola, Topics in the Theory of Riemann Surfaces. IX, 105 pages. 1994.

Vol. 1596: L. Heindorf, L. B. Shapiro, Nearly Projective Boolean Algebras. X, 202 pages. 1994.

Vol. 1597: B. Herzog, Kodaira-Spencer Maps in Local Algebra. XVII, 176 pages. 1994.

Vol. 1598: J. Berndt, F. Tricerri, L. Vanhecke, Generalized Heisenberg Groups and Damek-Ricci Harmonic Spaces. VIII, 125 pages. 1995.

Vol. 1599: K. Johannson, Topology and Combinatorics of 3-Manifolds. XVIII, 446 pages. 1995.

Vol. 1600: W. Narkiewicz, Polynomial Mappings. VII, 130 pages. 1995.

Vol. 1601: A. Pott, Finite Geometry and Character Theory. VII, 181 pages. 1995.

Vol. 1602: J. Winkelmann, The Classification of Three-dimensional Homogeneous Complex Manifolds. XI, 230 pages. 1995.

Vol. 1603: V. Ene, Real Functions – Current Topics. XIII, 310 pages. 1995.

Vol. 1604: A. Huber, Mixed Motives and their Realization in Derived Categories. XV, 207 pages. 1995.

Vol. 1605: L. B. Wahlbin, Superconvergence in Galerkin Finite Element Methods. XI, 166 pages. 1995.

Vol. 1606: P.-D. Liu, M. Qian, Smooth Ergodic Theory of Random Dynamical Systems. XI, 221 pages. 1995.

Vol. 1607: G. Schwarz, Hodge Decomposition – A Method for Solving Boundary Value Problems. VII, 155 pages. 1995.

Vol. 1608: P. Biane, R. Durrett, Lectures on Probability Theory. VII, 210 pages. 1995.

Vol. 1609: L. Arnold, C. Jones, K. Mischaikow, G. Raugel, Dynamical Systems. Montecatini Terme, 1994. Editor: R. Johnson. VIII, 329 pages. 1995.

Vol. 1610: A. S. Üstünel, An Introduction to Analysis on Wiener Space. X, 95 pages. 1995.

Vol. 1611: N. Knarr, Translation Planes. VI, 112 pages. 1995.

Vol. 1612: W. Kühnel, Tight Polyhedral Submanifolds and Tight Triangulations. VII, 122 pages. 1995.

Vol. 1613: J. Azéma, M. Emery, P. A. Meyer, M. Yor (Eds.), Séminaire de Probabilités XXIX. VI, 326 pages. 1995.

Vol. 1614: A. Koshelev, Regularity Problem for Quasilinear Elliptic and Parabolic Systems. XXI, 255 pages. 1995.

Vol. 1615: D. B. Massey, Lê Cycles and Hypersurface Singularities. XI, 131 pages. 1995.

Vol. 1616: I. Moerdijk, Classifying Spaces and Classifying Topoi. VII, 94 pages. 1995.

Vol. 1617: V. Yurinsky, Sums and Gaussian Vectors. XI, 305 pages. 1995.

Vol. 1618: G. Pisier, Similarity Problems and Completely Bounded Maps. VII, 156 pages. 1996.

Vol. 1619: E. Landvogt, A Compactification of the Bruhat-Tits Building. VII, 152 pages. 1996.

Vol. 1620: R. Donagi, B. Dubrovin, E. Frenkel, E. Previato, Integrable Systems and Quantum Groups. Montecatini Terme, 1993. Editors:M. Francaviglia, S. Greco. VIII, 488 pages. 1996.

Vol. 1621: H. Bass, M. V. Otero-Espinar, D. N. Rockmore, C. P. L. Tresser, Cyclic Renormalization and Auto-morphism Groups of Rooted Trees. XXI, 136 pages. 1996.

Vol. 1622: E. D. Farjoun, Cellular Spaces, Null Spaces and Homotopy Localization. XIV, 199 pages. 1996.

Vol. 1623: H.P. Yap, Total Colourings of Graphs. VIII, 131 pages. 1996.

Vol. 1624: V. Brînzănescu, Holomorphic Vector Bundles over Compact Complex Surfaces. X, 170 pages. 1996.

Vol.1625: S. Lang, Topics in Cohomology of Groups. VII, 226 pages. 1996.

Vol. 1626: J. Azéma, M. Emery, M. Yor (Eds.), Séminaire de Probabilités XXX. VIII, 382 pages. 1996.

Vol. 1627: C. Graham, Th. G. Kurtz, S. Méléard, Ph. E. Protter, M. Pulvirenti, D. Talay, Probabilistic Models for Nonlinear Partial Differential Equations. Montecatini Terme, 1995. Editors: D. Talay, L. Tubaro. X, 301 pages. 1996.

Vol. 1628: P.-H. Zieschang, An Algebraic Approach to Association Schemes. XII, 189 pages. 1996.

Vol. 1629: J. D. Moore, Lectures on Seiberg-Witten In-variants. VII, 105 pages. 1996.

Vol. 1630: D. Neuenschwander, Probabilities on the Heisenberg Group: Limit Theorems and Brownian Motion. VIII, 139 pages. 1996.

Vol. 1631: K. Nishioka, Mahler Functions and Trans-cendence.VIII, 185 pages.1996.

Vol. 1632: A. Kushkuley, Z. Balanov, Geometric Methods in Degree Theory for Equivariant Maps. VII, 136 pages. 1996.

Vol.1633: H. Aikawa, M. Essén, Potential Theory – Selected Topics. IX, 200 pages.1996.

Vol. 1634: J. Xu, Flat Covers of Modules. IX, 161 pages. 1996.

Vol. 1635: E. Hebey, Sobolev Spaces on Riemannian Manifolds. X, 116 pages. 1996.

Vol. 1636: M. A. Marshall, Spaces of Orderings and Ab-stract Real Spectra. VI, 190 pages. 1996.

Vol. 1637: B. Hunt, The Geometry of some special Arithmetic Quotients. XIII, 332 pages. 1996.

Vol. 1638: P. Vanhaecke, Integrable Systems in the realm of Algebraic Geometry. VIII, 218 pages. 1996.

Vol. 1639: K. Dekimpe, Almost-Bieberbach Groups: Affine and Polynomial Structures. X, 259 pages. 1996.

Vol. 1640: G. Boillat, C. M. Dafermos, P. D. Lax, T. P. Liu, Recent Mathematical Methods in Nonlinear Wave Propagation. Montecatini Terme, 1997. Editor: T. Ruggeri. VII, 142 pages. 1996.